Neues verkehrswissenschaftliches Journal

Ausgabe 24

Approach to Determine and Evaluate the Homogeneity of Operating Programs in Railway Operation Based on Infrastructure Occupancy

Von der Fakultät Bau- und Umweltingenieurwissenschaften der Universität Stuttgart
zur Erlangung der Würde einer Doktor-Ingenieurin (Dr.-Ing.)
genehmigte Abhandlung

Vorgelegt von
Nan Cao
aus Jiangxi, VR China

Hauptberichter: Prof. Dr.-Ing. Ullrich Martin
Mitberichter: Prof. Dr.-Ing. habil. Jürgen Siegmann

Tag der mündlichen Prüfung: 25/09/2017

Institut für Eisenbahn- und Verkehrswesen der Universität Stuttgart
2017

Titelbild: Nan Cao

Herstellung und Verlag: Books on Demand GmbH, Norderstedt

Printed in Germany

ISBN 978-3-7460-3326-6

Preface

Railway operation is normally based on very detailed timetables that strongly determine the operation process on a given infrastructure. There has always been a great effort to exploit the existing, costly infrastructure as much as possible early on in the timetable scheduling phase by maximizing its number of train runs, and at the same time by not reducing the quality of the train operations. Based on extensive empirical findings and the results of individual, topic-specific investigations, it is indisputable in this field of research that an operating program that is as homogeneous as possible can significantly contribute to good operation quality.

In this dissertation, a generic approach is developed for quantitatively determining the homogeneity of operating programs. This approach can be applied universally for different types and dimensions of investigated infrastructures, and also without consideration to technical conditions such as the dynamic parameters of train movements. This makes it possible to comprehensively and quantitatively describe the homogeneity of any given operating program directly using the operating program itself and its three parameters of homogeneity, which quantify the variation in blocking times, the variation in buffer times, and the variation in running directions. The homogeneity can be determined for individual occupancy elements as well as diverse infrastructure networks. A quantitative and generic approach for determining the homogeneity of operating programs with minimized effort (time and data) is developed in this dissertation, which is a considerable scientific contribution to the efficient planning and implementation of railway operations.

Stuttgart, December 2017

Ullrich Martin

Table of Content

List of Figures

List of Tables

Kurzfassung

Bei steigender Verkehrsnachfrage verbunden mit der Forderung der Sicherung einer bestimmten Betriebsqualität spielt die Struktur des Betriebsprogramms eine wesentliche Rolle für die effiziente Infrastrukturausnutzung. Die bisherige Forschung deutet darauf hin, dass die Infrastrukturausnutzung mit einem homogenen Betriebsprogramm effizienter ist. In den letzten Jahrzehnten versuchten mehrere Forscher, die Homogenität eines Betriebsprogramms im Eisenbahnverkehr zu beschreiben und / oder zu definieren; Jedoch keine Definition berücksichtigt die Belegung der Infrastruktur. Es gibt mehrere Methoden, um die Homogenität im Eisenbahnbetrieb unter Berücksichtigung von Geschwindigkeit, Fahrzeit und Zugfolgezeit zu bewerten. Aber die Belegung der Infrastruktur wurde in den bestehenden Methoden nicht diskutiert. Eine Erweiterung bestehender Definitionen wird aus der Perspektive der Belegung der Infrastruktur darstellen lässt, die sich durch Variationen in Belegungszeit, Pufferzeit und Fahrtrichtung auszeichnen kann. Die Belegungszeit, die Pufferzeit und die Fahrtrichtung beschreiben die Belegung der Belegungsabschnitte auf Basis der Sperrzeittreppe. Dementsprechend wird die Homogenität eines Betriebsprogramms durch drei Parameter ausgewertet, nämlich die Homogenität der Sperrzeit (HBL), die Homogenität der Pufferzeit (HBU) und die Homogenität der Fahrtrichtung (HRD). Die Gesamthomogenität (OH) kombiniert HBL; HBU und HRD zur integrativen der Bewertung der Homogenität von Betriebsprogrammen. Die Ergebnisse zeigen, dass die neue Methodik die Homogenität des Bahnbetriebes nicht nur für Belegungsabschnitte, sondern auch für ein ganzes Netzwerk genutzt werden kann, was wesentlich zur Effizienzsteigerung der Infrastrukturausnutzung beiträgt.

Mit der in dieser Dissertation entwickelten Methode wird die Wechselbeziehung zwischen der Homogenität eines Betriebsprogrammen und der Betriebsqualität quantitativ untersucht. Zunächst wird sowohl der Einfluss jedes homogenen Parameters als auch die Gesamthomogenität auf einer S-Bahn-Strecke analysiert. Die Betriebsqualität (Verspätungskoeffizient) wird für Fahrpläne mit unterschiedlichen Zugarten, Anordnungen von Pufferzeiten und Sequenzen von Zugfahrten ausgewertet. Die Ergebnisse zeigen, dass sich die Betriebsqualität mit weniger homogenen Fahrplänen verschlechtert, bei denen die Veränderung der Pufferzeit einen wesentlicheren Einfluss hat. Schließlich wird der Einfluss der Homogenität der Betriebsprogramme auf unterschiedliche Belastungen untersucht. Eine geringe Belastung ist empfindlicher gegenüber der Homogenität der Pufferzeit und die Homogenität der Sperrzeit ist für eine hohe Belastung signifikanter.

Abstract

Contemporary railway traffic requires different train services share the same infrastructure. Capacity consumption together with the homogeneous level of train traffic gives a picture of the efficient use of infrastructure. Previous research suggests that the utilization is more efficient with a homogeneous operating program. During last decades, many researchers tried to describe and/or define the homogeneity of operating programs in rail service; however, no definition considered the infrastructure occupation. In addition, several methods were developed to evaluate the homogeneity in railway operation considering variations in speed, running time and headway. But, the occupancy of infrastructure has not been discussed in the existing methods. An extension of existing definitions is presented from the perspective of the infrastructure, which can be characterized by variations in blocking time, buffer time and running direction. The blocking time, buffer time and running direction describe the occupancy of train path on track sections based on the blocking time model. Accordingly, the homogeneity of operating programs is evaluated through three parameters, namely the homogeneity of blocking time (HBL), the homogeneity of buffer time (HBU) and the homogeneity of running direction (HRD). The overall homogeneity (OH) combined HBL; HBU and HRD to realize an integrity evaluation of homogeneity of operating programs. The results show that this new methodology can quantify the homogeneity of railway operations, not only for track sections but for an entire network, which contributes significantly to the efficient utilization of infrastructure.

With the method developed in this thesis, the interrelationship between homogeneity of operating programs and operation quality were investigated quantitatively. Firstly, both the influence of each parameter of homogeneity and the overall homogeneity were analyzed. The operation quality (delay-coefficient) is evaluated for timetables with different train types, arrangements of buffer times and sequences of train runs. The results show that the operation quality deteriorates with less homogeneous timetables, in which the variation in buffer time has a more significant influence. Finally, the influence of homogeneity of operating programs is studied for different traffic flows. The low traffic flow is more sensitive to the homogeneity of buffer time and the homogeneity of blocking time is more significant for high traffic flow.

1 Introduction

Currently, worldwide railway operation confronts a general increase of traffic demand along with the economic development. In many regions, the railway operation is approaching the capacity limits[1]. At the same time, the operation quality will be worse with the increasing traffic flow in the form that the railway operation is more sensitive to disturbances, and thus results in larger delays and lower average punctuality. On the other hand, different traffic demands require diversification have led to capacity shortage and unsatisfied operation quality in railway network [Lindfeldt 2010]. Therefore, how to accommodate this significant amount of mixed traffic demand with adequate operation quality is a major issue for infrastructure manager as well as train operators.

This issue can be addressed through upgrading infrastructure such as doubling main railway tracks, extending railway stations, constructing new lines or junctions, as well as upgrading existing infrastructure to provide new capacity or improve operation quality [Landex 2008; Lüthi 2009]. Current infrastructure upgrading mainly relies on experienced capacity planners and simulation software to identify bottlenecks, analyze causes and plan upgrades through long-term strategic capacity planning projects [Lai and Barkan 2011; Martin and Li 2014]. In addition, upgrading of signaling systems reduces average headways, which also enhances an efficient management of train runs [Eichenberger 2007]. Apparently, infrastructure improvement has significant effects, while it is inarguably not an efficient solution regarding money, time and labor. In addition, existing railway infrastructure is not easy to adjust or change, particularly in bottlenecks and high-density station areas. The space for an extensive construction is limited after decades of infrastructure development, especially in urban railway systems. Compared with infrastructure investment, high efficient utilization of existing infrastructure turns out to be a flexible solution that is more cost-efficient and quick-implemented to deal with the requirements of increasing capacity and/or better operation quality, especially for medium and short term capacity planning [Lai and Barkan 2011].

Efficient use of infrastructure requires a certain amount of train paths; on the other hand, the operation quality should meet the requirement of passengers and operators. The key factor in the utilization of infrastructure is the interaction of different train types. Schittenhelm points

---

[1] Capacity Limit: A typical value that if the infrastructure occupation [% of time-window] is higher than or equal to the analyzed line section shall then be called congested infrastructure and no more additional train paths may be added to the timetable (UIC 2013).

out that capacity consumption levels compared with homogeneity level of train traffic can give a picture of efficient use of the available infrastructure [Schittenhelm 2014].

Previous work has investigated the effects of inhomogeneous traffic on the utilization of infrastructure in railway systems. Bronzini and Clarke [Bronzini and Clarke 1985] used a single-track simulation model to compare the delay-volume curves of different mixtures of intermodal and unit trains. Vromans et al. [Vromans et al. 2006] studied the effect of the inhomogeneity of various passenger service on robustness of timetable. Abril et al. [Abril et al. 2008]and Landex et al [Landex and Jensen 2013] also analyzed the capacity with heterogeneous traffic. The train characteristics of speed, acceleration, braking, and priority were considered by Dingler et al. [Dingler et al. 2014] for their effect on the increased delays due to heterogeneity. Anders Lindfeldt [Lindfeldt 2015] simulated several hundred scenarios to analyze the influence of traffic heterogeneity on knock-on delays, used timetable allowance and capacity. It is generally accepted that an efficient utilization of existing infrastructure can be achieved with a more homogeneous operation.

From another perspective, inhomogeneity as the results of offering customized service level for various market segments usually leads to larger capacity consumption and smaller buffer times between consecutive trains (smaller buffer times normally located where trains catch up each other) if no improvement has been made to signaling system and infrastructure, which may increase delay propagations in operation [Abril et al. 2008; DB Netz AG 2008; UIC 2013; Pachl 2013; Schittenhelm 2014]. Therefore, understanding how the traffic mix affects capacity and quality is necessary to use capacity efficiently.

It is not easy to precisely define homogeneity in railway operation that many researchers tried to address this problem for last decades [Vromans et al. 2006; Landex 2008; UIC 2004; Lindfeldt 2013; Schittenhelm 2014]. In UIC leaflet 406 [UIC 2004], inhomogeneity is related to differences in running times between different train types on the same track. Vromans et al. considered railway traffic to be homogeneous if all trains have similar characteristics; especially the same average speed per track segment, resulting from the running times and the stopping times [Vromans et al. 2006]. Another description, based on timetables, is given by Landex who explained that a timetable is homogeneous when there is no variation in the speed, the stop pattern, and the headway times for a line section [Landex 2008]. According to Anders Lindfeldt [Lindfeldt 2013], heterogeneity can be used to describe two different properties of the timetable. The first is how evenly distributed the train movements are over a given period of time and the second is associated with speed differences between trains.

In general, the variations in the speed (and thereby the stop pattern) and/or the headways were taken into account to describe the homogeneity. No literature, however, has been found considering the occupation of infrastructure so far. Only a DB guideline classifies the operating program based on the state of infrastructure occupation [DB Netz AG 2008]. It divided the operating program into homogeneous operating program and inhomogeneous operating program (also named mixed operating program). In a homogeneous operating program, the trains operate with nearly same occupancy time, e.g. in commuter rail or high-speed rail network. Alternately, trains with different occupancy time are operated in inhomogeneous operating program. However, no further research has been done based on this understanding. Therefore, a new definition of homogeneity is presented in this article from the standpoint of occupancy of infrastructure based on the description of [DB Netz AG 2008] to enrich the understanding of homogeneity in railway systems.

Based on these concepts of homogeneity, there are various approaches to evaluating homogeneity quantitatively. Hertel put forward that the coefficient of variation of inter-arrival time[2] (headway) and service time[3] (minimum headway) indicate the homogeneity in railway operation which determine the lower and upper limitations of the "recommended area of traffic flow[4]" [Hertel 1995; Chu 2014]. In 1999 Carey described several methods to measuring the homogeneity focused on headways between consecutive trains [Carey 1999]: the percentage of headways smaller than a certain size; the percentiles of headways distribution; and the range, the standard deviation, the variance or the mean absolute deviation of headways. The headways were measured at single locations without the behavior of the trains on the surrounding track section in these approaches. Therefore, measurement indicator SSHR (Sum of Shortest Headways Reciprocals) considering the smallest headways between two consecutive trains on a particular track section instead of at one single location was developed by Vromans et al. in 2006 [Vromans et al. 2006].

---

[2] Inter-arrival time: The inter-arrival time is the difference between succeeding arrival times of requested trains or train paths in queueing theory, which is equivalent to the headways in railway operation [Hertel 1992, Chu 2014].

[3] Service time: Service time is the time a customer (train) occupies a certain part of the infrastructure for service based on the queueing theory. In railway operation, the service times are often represented by the minimum headway times (not to be confused with blocking time) [Hertel 1992, Chu 2014].

[4] The recommended area of traffic flow defines the traffic flow, in which traffic operates economically with satisfied operation quality. The lower limit is given by the minimum value of the relative sensitivity function of the waiting times. The upper limit is the maximum value of the so-called traffic energy function [Hertel 1992, Chu 2014].

With the assumption that the headway at arrival station is more important than that at departure station for knock-on delays, indicator SAHR (Sum of Arrival Headways Reciprocals) was further established considering only the headways at arrival station. If the running time of each train and the order of trains running are pre-specified, the railway operation reaches a homogeneous situation when the minimum headways between subsequent trains are equalized during the whole time spread. Measurements were further improved taking both the variation in headways and speeds into account [Landex 2008; Lindfeldt 2015]. The SAHR is equal to the SSHR only when the trains are running with same speed. In inhomogeneous cases, the SAHR is smaller than the SSHR, and the difference between them increases with the differences in speed. Accordingly, the ratio of SAHR and SSHR was calculated as an indicator to reflect the variations in the speed [Landex 2008]. Nevertheless, this ratio is still equal to 1 if the identical trains are operated with different headway times, which is not homogenous operation. Therefore, the ratio of the headway at departure station to the following headway multiplied by the ratio of headways for arrival at stations took the variations of headways into account. The variations in speed were implicitly taken into accounts as varying speed resulted in variation in the headway times at departure station and arrival station.

Indicators MDSR (Mean of the Difference in Scheduled Running time) and MDFR (Mean of the Difference in Free Running time) were presented as supplements to SSHR and SAHR in order to describe the differences in running time-based on the timetable [Lindfeldt 2015]. MDFR was found to have higher explanatory value due to that it is independent of traffic load, while MDSR depends on the amount of traffic load. In summary, these methods utilize the headways at stations and/or running time through the whole section. It is difficult for these methods to obtain the overview of the homogeneity of operation in the entire network. In addition, the existing methods do not fit the new definition of homogeneity considering the infrastructure occupation. A new occupancy based method of homogeneity evaluation is consequently developed in this thesis, which enables the evaluation of homogeneity in a network based on the new definition considering the occupancy of infrastructure.

The importance of understanding homogeneity increases when more and more trains are operated on the given infrastructure. It is because of that more trains on the infrastructure in general result in a larger amount of possible conflicts between the trains, especially for train mix. A better understanding of homogeneity becomes an important issue for improving the utilization efficient of existing infrastructure. It also makes it possible to adjust the structure of an operating program in order to optimize both the capacity and quality in operation. Therefore, a comprehensive investigation related to the homogeneity of operating programs conducted in author's research focus on the infrastructure occupation. The aim and structure of the thesis are outlined here.

- **Objective**

The main objective of the thesis is to identify and evaluate the homogeneity of operating programs from the aspect of occupancy of infrastructure and to find out how the homogeneity affects railway operation in railway systems.

Firstly, a systematical description of homogeneity of operating program in railway systems should be developed from the aspect of infrastructure occupation. Homogeneity involves a lot of factors, and it is difficult to identify. Therefore, understanding the underlying principles of railway operation and how they affected homogeneity is important. In last decades, a lot of research has been done related to this issue. The homogeneity is related to speed, running time, stop pattern and headways according to these researches. However, according to author's best knowledge, no literature considers the occupancy of infrastructure for homogeneity. A new definition of homogeneity considering the occupancy of infrastructure should be introduced to enrich the current understanding of homogeneity. The new occupancy based definition is significant for infrastructure manager to use the existing infrastructure by homogenizing train runs.

To be able to quantitative analyze the influence of homogeneity of operating program; it is necessary to evaluate the homogeneity quantitatively. The evaluation method is to extract enough meaningful information from abundant data to reflect he homogeneous level. Existing methods don not meet the new definition very well. Another objective of the thesis is to generate an integrated evaluation system for homogeneity from the aspect of infrastructure occupation. The developed method should not only be able to evaluate the homogeneity for a single track section but also implement an overall evaluation of homogeneity in the entire investigated network.

The operation quality of is checked for operating programs with different levels of homogeneity based on a new evaluation system of homogeneity. Taking a commuter rail as an example, the interrelationship between the homogeneity and the operation quality will be analyzed quantitatively. The relationship between homogeneity and quality of operation is important to improve the robustness of timetables. Additionally, it provides timetable planners a good principle in timetable scheduling process.

The homogeneity of operating programs is also of great importance in capacity research of mixed traffic. Investigating the influence of homogeneity by increased traffic flow is another objective of the thesis. The results can be used in practical operation to find out a homogeneous structure of operating program in order to operate more trains on the infrastructure with satisfied operation quality. It is quite meaningful for capacity managers to use the existing infrastructure with good operation quality. This study is especially significant for a congested railway station in the urban area, for which the possibility to extend or to upgrade the existing infrastructure is small.

- **Structure**

This thesis is divided into three parts. The first part (Chapter 1 and 2) gives an introduction of homogeneity of operating program in railway operation. It points out the importance of having sufficient knowledge about homogeneity and how, over time, it becomes possible to increase the efficiency of infrastructure utilization by improving the arrangement of trains with the growth of traffic flow. In the second part (Chapter 3), a new occupancy based definition of homogeneity is developed based on the blocking time model, which considers the infrastructure occupation. Accordingly, a new occupancy based method is presented to evaluate the homogeneity quantitatively. This occupancy based method is also compared with other methods. The third section (Chapter 4, 5 and 6) uses the occupancy based method to quantify the homogeneity of operating program and analyze the influence of homogeneity on operation quality. The effect of homogeneity in capacity research is also investigated.

Chapter 2 introduced the basic principles of railway operation, in which operating program is a critical and flexible component for railway operation. Operating program[5] arranges different types of trains operated on the infrastructure. The operating program parameters[6] are classified into four classes. The structure of operating program determines the capacity and operation quality in railway operation. Homogeneous operating program has better operation quality comparing to the corresponding inhomogeneous operating program. Additionally, the research related to the homogeneity of operating programs is presented in this chapter. The existing definitions and evaluation methods are introduced in detail.

Chapter 3 presents a new understanding of homogeneity of operating program from the aspect of infrastructure occupation. This chapter starts with an introduction of the blocking time

---

[5] Operating program: In railway system, the operating program is the comprehensive date-related description of the performance and requirements of railway operation, including amounts of train runs, properties of trains, structure and sequence, as well as temporal allocation of the train runs.

[6] Operating program parameter: The characteristic is to be considered in the operating program for the performance of railway operation in order to achieve the traffic demand.

model, which describes the occupation of a train path on the network. Based on this model, the influence of various factors on the homogeneity of operating programs is systematically analyzed. A new definition of homogeneity of operating program in railway systems is consequently developed through a new perspective of infrastructure occupation, considers the variations in blocking time, buffer time and running time. This definition extends and completes the present understanding of homogeneity, which is more significant for efficient infrastructure utilization.

However, existing measures of homogeneity are not compatible with the new occupancy based definition, hence; a different measurement based on the occupancy based definition of homogeneity of operating program is proposed and discussed in detail in the next step. The method starts to evaluate the variations in blocking time, buffer time and running direction for a single track section. In order to assess the homogeneity in the whole network, the variations on each track section are weighted. The deduction process will be introduced in Chapter 3.This occupancy based method is able to measure the homogeneity of a track section, but also for a network. The feasibility of this occupancy based method and the comparison of existing methods and the occupancy based method are also presented.

Based on the new occupancy based method of homogeneity evaluation, a study regarding the interrelationships between operation quality and parameters of homogeneity are carried out in Chapter 4. The case study is based on the real commuter rail system in Germany which is able to operate entire homogeneous operating program and inhomogeneous operating program. The influence of homogeneity is checked for each parameter of homogeneity.

Up to now, the homogeneity is evaluated separately for each parameter of homogeneity, e.g. homogeneity of blocking time, homogeneity of buffer time and homogeneity of running direction. In Chapter 5, these three parameters of homogeneity are integrated to create an overall homogeneity based on the weighted Euclidean distance of three parameters of homogeneity. The entropy of a parameter of homogeneity determines its weight to the overall homogeneity. The overall homogeneity evaluation models are developed for both a commuter rail network and a mixed traffic network. The influence of this overall homogeneity is also investigated for these two networks.

The capacity research investigates the relationship between traffic flow and operation quality within the studied railway system. The relationship is also related to the structure of operating program. In Chapter 6, the influence of homogeneity in capacity research is analyzed based on several different scenarios of operating program. Finally, a conclusion and remarks to future research will be provided in Chapter 7.

2 Basic Principles in Railway Operation

2.1 Essential Components in Railway Operation

Railway operation is a product of complex interaction of different components. Generally, the following 4 components have to be considered in railway operation:

- The infrastructure with the layout of tracks, the signaling equipment, the stations and on electrified lines, the catenary or third rail system with power supply.
- The rolling stock with cars and locomotives, which operates on the infrastructure.
- The operating program, which arranges the movements of these rolling stocks on the infrastructure.
- The rules and procedures in railway operation for a safe and efficient operation.

Tracks are the roadways of a railway system, on which operate the rolling stocks. The dimension of the infrastructure has a major impact on capacity. Auxiliary tracks at crossing stations increase the capacity of a single track because they enable trains to carry out crossing and overtaking avoiding meets. According to the work of Transportation Research Board in Washington and Abril et al., double tracks usually have greater than twice capacity of a single track [Abril et al. 2008; Transportation Research Board 2013]. However, a four-track line rarely increases capacity by more than 50% over a double-track line. Apparently, stations are the most influential components for railway operation. In Europe, busy complex railway stations are the key components of passenger transport, since they are the locations of most conflicts. Normally, the station is a crucial bottleneck in railway network.

The signal system guarantees a safe distance between trains. There are two typical types of systems: fixed block signaling systems and moving block signaling systems [Pachl 2015]. In the fixed block signaling systems, trains are separated by fixed block sections which are protected by signals. The signals provide trains the movement authority to enter the block section. In a moving block signaling system, the location data of trains is transmitted continuously at short intervals. This requires an efficient communication system between line signals, cabs and control centers. In the future, the ETCS[7] specification contains a Level 3 which is based on moving block with respect to future applications.

Operation rules thus impact the design and construction of both infrastructures and rolling stocks and they decisively affect the rail system's performance and efficiency. Depending on technical or regulatory developments, the operation rules are adjusted consistently to enhance the effectiveness and competitiveness of rail traffic. The priorities of trains play a vital role. Train priorities decrease capacity because priority trains are given preferential treatment over lower priority trains, which results in increased delays. This allows the high priority traffic to move as if it were the only traffic in the network. Generally speaking, the greater the number of priority classes, the less capacity is available.

In most railway operation, different rolling stocks share the same infrastructure simultaneously. The train characteristics, such as speed, acceleration and breaking behavior, are important elements in estimating running time [Pachl 2015]. Higher speed on one hand reduces the occupation time of the block section, but increases the breaking distance on the other hand. In last decades, several countries developed their own high speed trains profits from the modern scientific and technological development. The French TGV[8] set a speed record as 574 km/h in 2007. As well, the German ICE[9] reached the maximum speed as 406 km/h, and the high speed train can reach the top speed as 300 km/h in China. The operation of these high speed trains has put forward the higher requirements of infrastructures and signal systems.

The arrangement of these train movements is made through an operating program, in which train mix is an important attribute. The ideal situation is that identical trains operate in the network. However, current railway operation requires different train services with different origins and destinations and various halting patterns. As the mixture of different trains increases, more conflicts are generated. This is probably a main reason for the delay propagation in the railway network. Therefore, how to accommodate such significant amount of mixed traffic demands with satisfied operation quality is a major issue for infrastructure managers as well as train operators.

---

[7] ETCS (European Train Control System) is the signaling and control system of the European Rail Traffic Management System.

[8] TGV is intercity high-speed rail service operated by the national rail operator in France.

[9] ICE is a system of high-speed trains predominantly running in Germany offered by Deutsche Bahn.

Nowadays, infrastructure, rolling stocks, operating program, as well as operation rules and procedures are designed together in a way to transport passengers and goods to the destinations safely, quickly, punctually.

2.2 Structure of Operating Program

2.2.1 Classification of Operating Program Parameters

In railway systems, the operating program is a comprehensive date-related description of the performance and requirement of the railway operation. According to DB NETZ guideline 405 [DB Netz AG 2008], an operating program is related, not only to characteristics of rolling stocks themselves but also to the structure of train runs, including the amount of train, the properties of trains, the structure and sequence of train runs, as well as temporal allocation of train runs.

Depending on task requirement, time horizon, method, and tool, the required or possible levels of specification for the operating program is different. The operating program can be detailed as a fully scheduled timetable for passengers, or it also can be general information, such as the number of trains needed for each line per day. Timetables are usually available only for short-term time horizons. For medium and long term planning, appropriate modifications must be made.

The efficient utilization of a railway network is highly dependent on the operating program. This structure generalizes the potential influencing operating program parameters and classifies these parameters into four categories based on a thorough review of the literature:

- **Train-related operating program parameters** focus on the attributes of trains; including their configuration characteristics and dynamic behaviors. The train length, which has an important influence on occupation time, belongs to the configuration characteristics; meanwhile, maximum speed, accelerating, and deceleration behavior are part of the dynamic behaviors.
- The supply and frequency of train lines would be determined at the initial stage of scheduling to guarantee the demand for railway traffic. For each train line, the exact blocking time should be determined, for which the recovery time and dwell time reserve are critically influential. The characteristics involved in the operational use of a line by trains are treated as **line-related operating program parameters**.
- Homogeneity in railway systems embodies itself into two aspects. One is the different running time due to various train-related and line-related operating

program parameters. The other is the structure of trains running which is defined as **structure-related operating program parameters**. In practice, headways are assigned based on the combination of consecutive train movements. In mixed traffic railway network, average running time is related to share of train runs; the sequence affects the headway between successive trains.

- In the real operation, the priority of different trains and corresponding dispatching rules determine the operation quality; these are defined as **dispatching-related operating program parameters**. Some results have been found based on DFG10 project (MA 2326/15-1) concerning the influence of minimum lateness for dispatching in capacity research [Liang et al. 2015; Martin and Liang 2017].

2.2.2 Influence Factors of Operating Program

A key factor of the operating program affecting the usage of infrastructure comes from different train types running in the network. These trains have substantially different configuration profiles and dynamic characteristics, which are widely proved possessing great influence on capacity and operation quality [Abril et al. 2008; Pachl 2013]. Dingler investigated the impact of train-related parameters, including speed, braking, acceleration and priority [Dingler et al. 2014].

In addition, the inhomogeneity resulting from mixture and order of train running also has a negative influence on the capacity and the operation quality of a railway system [Lai and Barkan 2011; Pachl 2013]. Bronzini and Clarke used a single-track simulation model to compare the delay-volume curves regarding different mixtures of intermodal and unit trains [Bronzini and Clarke 1985].The influence of inhomogeneity on railway operation is evaluated by two indicators regarding headways between following trains in [Vromans et al. 2006].

The operation frequencies of different trains are systematically treated as timetable variables in [Yuan and Hansen 2007] and [Lindfeldt 2010] to evaluate the influence coming from timetables on railway operation. Other characteristics in timetable construction, such as recovery time, dwell time reverse, are analyzed in [Carey 1999] and (Summary Report: Capacity Management (Capman Phase 3), 2004)[UIC 2004].

---

[10] DFG (German: Deutsche Forschungsgemeinschaft, English: German Research Foundation) is a German research funding organization.

Dispatching strategy is also a vital factor, which is investigated by an ongoing DFG-Project (MA 2326/15-1) at IEV[11] [Martin and Liang 2017]. The research proved that the efficient utilization of infrastructure is connected to these operating program parameters. However, with respect to the topic, the analysis of operating program parameters is only partially investigated and not in an overall comprehensive systematic manner.

Furthermore, few studies have been conducted to examine the combined influence of multiple operating program parameters on homogeneity. Krueger studied the influence of operating program parameters, including speed ratio and average speed, with different traffic volume on train delays [Krueger 1999]. The combined influence of capacity consumption and recovery time on delays is investigated in (Summary Report: Capacity Management (Capman Phase 3), 2004)[UIC 2004].

However, utilization of existing infrastructure is always related to several operating program parameters. For instance, replacing slower rolling stocks changes the structure of the trains which meanwhile increase their average speed. In addition, strong interactions between operating program parameters are expected. Hence, it is interesting to analyze the combined influence of multiple operating program parameters in consideration of the interactions between the operating program parameters. There seems to be no well-known published literature related to a systematic and comprehensive analysis of operating program parameters and their relation to a universal definition of homogeneity at present.

2.3 Homogeneity in Railway Operation

An efficient utilization of the existing infrastructure is an essential component of a high-quality transportation system and has become a crucial task for railway infrastructure management. One solution to increase the reliability is to reduce the propagation of delays due to the interdependencies between trains. Reduction of the running time differences per track section, i.e. by generating more homogeneous timetables is one way to reduce the propagation of delays.

2.3.1 Existing Definitions of Homogeneity

As mentioned above, the structure of operating program plays an important role in efficient use existing infrastructure, influencing the capacity and operation quality in railway operation. It is accepted that better operation quality can be achieved with a more homogeneous operating program. During last decades, researchers have put hard efforts related to this issue,

---

[11] IEV (Institut für **E**isenbahn- und **V**erkehrswesen): Institute of Railway and Transport Engineering in the university of Stuttgart.

attempting to identify the homogeneous railway operation. Over the years, homogeneity has been defined in different ways in the literature.

In UIC leaflet 406 [UIC 2004], inhomogeneity is related to differences in running times between different train types on the same track. When the differences in running time between different kinds of train worked on the same track are marked; similarly, the capacity consumption of the same number of trains will increase proportionately.

Vromans et al. considered railway traffic to be homogeneous if all trains have similar characteristics; especially the same average speed per track segment, resulting from the running times and the stopping times [Vromans et al. 2006]. Appropriate examples of homogeneous railway traffic are metro systems where all trains have the same running times per track and where all trains stop at all stations. If there are significant differences in the timetable characteristics of the trains on the same track, then the railway traffic is called heterogeneous, such as national railway networks with great differentiation in passenger services.

Another description, based on properties of timetables, is given by Landex who explained that a timetable is homogeneous when there is no variation in the speed, the stop pattern, and the headway times for a line section [Landex 2008]. In a homogeneous timetable, a train would not catch up with one another. But, in an inhomogeneous timetable, a fast train would catch up a slow train if not enough headway is assigned. The deceleration and acceleration process due to scheduled stop also influence the running time. Therefore, the stop pattern is also considered by Landex to identify homogeneity.

According to Anders Lindfeldt [Lindfeldt 2013, Lindfeldt 2015], heterogeneity can be used to describe two different properties of the timetable. The first is how evenly distributed the train movements are over a given period of time and the second is associated with speed differences between trains. In heterogeneous timetables, trains use the infrastructure unevenly over time with a distinct difference in average speed. On the contrary, homogeneous timetables arrange the trains at equal speed evenly distributed over time.

In general, the variations in the speed (and thereby the stop pattern) and/or headways were taken into account to describe the homogeneity in railway systems. However, the speed and headways cannot reflect the status that infrastructure is occupied by train runs, which is few considered in the literature.

According to [DB Netz AG 2008], the operating program can be classified into two types: homogeneous operating program and inhomogeneous operating program (mixed operating program). In a homogeneous operating program, the trains operate with almost same occupancy time, e.g. in commuter rails (S-Bahn in Europe) or some high-speed rail networks. Alternately, trains with different occupancy time are operated with an inhomogeneous operating program. This definition of homogeneity considering the occupation of infrastructure, but no further research is conducted based on this definition.

2.3.2 Existing Methods to Evaluate Homogeneity

Based on the existing definitions of homogeneity, there are various approaches to evaluating homogeneity quantitatively. The easiest way is using train mix to present the homogeneity in operation. Along decades, some indicators are developed to assess the homogeneity of operating program. Here important approaches are introduced in detail.

- **Statistical Analysis of Headways**

In 1999, Carey described several meaningful indicators at a station to reflect the homogeneity of operating program concerning the headway spread between consecutive trains [Carey 1999]. These indicators include:

- the percentage of headways smaller than a certain size,

- the percentiles of headways distribution,

- the range, the standard deviation, the variance and the mean absolute deviation of headways.

These approaches based on the principle that equalizing scheduled headways for one station has a positive influence on punctuality when the disturbance distributions are same for all trains. However, they didn't account the behavior of trains on the surrounding track sections since the headways are measured at one single location, and not on one track section.

- **SSHR, SAHR, and SSBR**

Vromans [Vromans 2005] proposed some heterogeneity indices that can measure the distribution and heterogeneity of trains over a given period. Two indicators **SSHR** (Sum of Shortest Headways Reciprocals) and **SAHR** (Sum of Arrival Headways Reciprocals) are developed to evaluate the homogeneity of operating program for line section during one cycle time. These two indicators are widely known in the evaluation of homogeneity.

SSHR considers the smallest headways between two consecutive trains on a particular track section instead of at one single location. It is a function of the smallest headways between two trains. For N trains per cycle, it is calculated with following formula:

$$SSHR = \sum_{i=1}^{n} \frac{1}{h_i^-}$$ Formula(1)

Where,

h_i^- is the smallest scheduled headway between train i and $i+1$ on the track section;

n is the number of trains in a cycle time.

The more homogeneous an operating program is, the smaller the value of SSHR. The SSHR is not only capable of representing the distribution of trains over the hour on a track section, but also of including the heterogeneity of these trains on this track. If the running time of each train and the order of trains running are pre-specified already, the operation reaches a homogeneous situation when the minimum headways between subsequent trains are equalized during the whole time spread.

The Sum of Arrival Headway Reciprocals (SAHR) considers only the headways at arrival point of the line section instead of the minimum headway time under the assumptions that for knock-on delays, headway at arrival is more important than at departure. SAHR is calculated with following formula:

$$SAHR = \sum_{i=1}^{n} \frac{1}{h_i^A}$$ Formula (2)

Where,

h_i^A is the headway at arrival point between train i and $i+1$ on the track section;

n is the number of trains in a cycle time.

The value of SSHR also lessens when the structure of operating program is more homogeneous with evenly spread headways on arrival point. As yet SAHR does not take the track into account anymore and is a single location measure in fact. An improved measure may be attained by taking the weighted average of the two measures above.

The heterogeneity measure SSHR and SAHR were discussed without referring to the technically minimal headway. However, during the operations, it does not matter what the absolute size of headway is, but what the buffer in this headway is: the difference between the planned headway and the minimum headway. Therefore, using buffers instead of headways for SSHR would lead to the **SSBR** (Sum of Shortest Buffer Reciprocals) with the following formula:

$$SSBR = \sum_{i=1}^{n} \frac{1}{(h_i - h_i^{min})^-} = \sum_{i=1}^{n} \frac{1}{b_i^-} \qquad \text{Formula (3)}$$

Where,

$(h_i - h_i^{min})$ is the smallest difference between the planned headway and the minimum headway between train i and $i + 1$.

These measures give a clear distinction between homogeneous timetables and heterogeneous timetables. The research by Vromans suggests that reducing the value of SSHR and SAHR, and thus increasing homogeneity, increases robustness as the risk of delay propagation from smaller initial delays is decreased. However, the value of SAHR and SSHR depends on the amount of traffic flow. They are not absolute measures, but are mainly meant to be able to compare different timetables for the same track or as in indication of how to produce a reliable timetable for a certain track.

- **Homogeneity Rate**

Measurements were further improved by taking the variation in headways and speeds into account in [Landex 2008]. In the homogeneous cases, SAHR is equal to SSHR. In inhomogeneous situation, meanwhile, SAHR is smaller than SSHR. A measurement of the homogeneity of the structure of operating program combining SSHR and SAHR was developed by Landex:

$$Homogeneity = \frac{SAHR}{SSHR} = \frac{\sum_{i=1}^{n} \frac{1}{h_i^A}}{\sum_{i=1}^{n} \frac{1}{h_i^-}} \qquad \text{Formula (4)}$$

Where,

h_i^- is the smallest scheduled headway between train i and $i + 1$ on the track section;

h_i^A is the headway at arrival point between train i and $i + 1$ on the track section. Train n is followed by train 1, due to cyclist;

n is the number of trains in a cycle time.

The SAHR is equal to the SSHR only when trains are running at same speed. In inhomogeneous cases, the SAHR is smaller than the SSHR, and the difference between them increases with the differences in speed. The ratio of SAHR and SSHR, therefore, was calculated as an indicator to represent the variation in the speed. Nevertheless, the calculated ratio is still equal to 1 if identical trains are operated with different headway times, resulting in an inhomogeneous operation.

Therefore, some methods are utilized aiming to avoid the difference misunderstanding. Landex proposed another measurement of homogeneity degree of operating program that uses the ratio of the headway at departure station ($h_{t,i}^{D}$) to the following headway ($h_{t,i+1}^{D}$) multiplied by the ratio of headways for arrival at stations ($h_{t,i}^{A}$, and $h_{t,i+1}^{A}$). To provide a formula independent of the number of trains, the result is divided by the number of headways minus $1(h_{N-1})$.

$$Homogeneity = \frac{\sum \left(min \left(\frac{h_{t,i}^{D}}{h_{t,i+1}^{D}} ; \frac{h_{t,i+1}^{D}}{h_{t,i}^{D}} \right), min \left(\frac{h_{t,i}^{A}}{h_{t,i+1}^{A}} ; \frac{h_{t,i+1}^{A}}{h_{t,i}^{A}} \right) \right)}{h_{N-1}} \qquad \text{Formula (5)}$$

The departure part and the arrival part include the variation of headways. The variation in speed is implicitly considered as varying speed results in variation in the headway times between the departure station and arrival station. The homogeneity indicator gives 1 when both the speed and the distribution of headways are equal. When the timetable becomes more inhomogeneous, the homogeneity indicator tends toward 0. These measures have an advantage that they are independent of traffic density and number of trains used in the calculation.

- **Hom$_A$ and Hom$_D$**

In [Landex and Jensen 2013], a homogeneity index Hom_A describing the train arrivals to a station and a homogeneity index Hom_D depicting the train departures from a station have been developed.

For the homogeneity of arrivals, it is the sum of the ratios between the arrival headway and the following arrival headway and divided it by the number of headways minus 1.

$$Hom_A = \frac{\sum min \left(\frac{h_{t,i}^{A}}{h_{t,i+1}^{A}} ; \frac{h_{t,i+1}^{A}}{h_{t,i}^{A}} \right)}{h_N - 1} \qquad \text{Formula (6)}$$

Where,

$h_{t,i}^A$ is the arrival headway between train i and $i + 1$ at station;

h_N is the number of headways.

Similarly, the homogeneity of the departures from the station can be calculated as the ratio between the headways time and the following headway time of the departures instead.

$$Hom_D = \frac{\sum min\left(\frac{h_{t,i}^D}{h_{t,i+1}^D}; \frac{h_{t,i+1}^D}{h_{t,i}^D}\right)}{h_N - 1}$$

Formula (7)

Where,

$h_{t,i}^D$ is the departure headway between train i and $i + 1$ at station;

h_N is the number of headways.

The homogeneity of departures from the station can be calculated using the ratio between the headway times and the headway time of the succeeding headway of the departure instead. These homogeneities can be calculated for the entire station to give an overview of arrival and departure pattern.

- **MDSR and MDFR**

Besides the influence of different headways, two measures **MDSR** (Mean of the Difference in Scheduled Running time) and **MDFR** (Mean of the Difference in Free Running time) were presented by [Lindfeldt 2015] to describe the differences in running time based on the timetable.

MDSR uses the total running time from the scheduled timetable, in which extra stops for overtaking are also included:

$$MDSR = \sum_{i=1}^{n} \frac{dsr}{n} = \sum_{i=1}^{n-1} \sum_{j=i+1}^{n} \frac{|srt_j - srt_i|}{n}$$

Formula (8)

Where,

dsr is the difference in scheduled running time between train i and $i + 1$ on the whole line;

srt_i is the scheduled running time of the ith train.

For MDFR, free running time means running times in the timetable without extra stops, as if trains are scheduled alone on the track. The difference is calculated for all combinations of trains in a cycle, not only between consecutive trains as if all trains in the cycle are likely to affect each other:

$$MDFR = \sum_{i=1}^{n} \frac{dfr}{n} = \sum_{i=1}^{n-1} \sum_{j=i+1}^{n} \frac{\left|frt_j - frt_i\right|}{n} \qquad \text{Formula (9)}$$

Where,

dfr is the difference in free running time between train i and $i+1$ on the whole line;

frt_i is the free running time of the ith train.

These two measures both involved the difference in running time of trains. With a homogeneous operating program, where trains run at the same speed, MDSR and MDFR have a value of 0. However, MDFR turns out to have higher explanatory value due to that it is independent of traffic lads, while MDSR depends on the traffic load.

Table 2-1: Summary of approaches to evaluate homogeneity of an operating program

	Indicators	Parameters	Consideration of Surrounding tracks	Independence of traffic flow	Evaluation Of a network
Carey [1999]	1. Percentage of headways smaller than a certain size 2. Percentiles of headways distribution 3. Range, var, s.d., or m.a.d of headways	Headways	✗	✓	✗
Vromans [2005]	SSHR	Headways	✓	✗	✗
	SSBR		✓	✗	✗
	SAHR*		✗	✗	✗
Landex [2008]	SAHR/SSHR**	Speed	✓	✓	✗
	Homo	Speed & Headways	✓	✓	✗
Lars W. Jensen [2013]	Hom$_A$	Headways	✓	✓	✗
	Hom$_D$		✓	✓	✗
Lindfeldt [2015]	MDSR	Running time	✓	✗	✗
	MDFR		✓	✓	✗

*: this indicator is calculated based on the assumption that the headway at arrival station is more important than that at departure station.

**: the ratio is still equal to 1 (means homogeneous operation) if identical trains are operated with different headway times.

These methods evaluate the homogeneity of operation at stations and/or on particular track sections, while it is hard for them to provide the overview of the homogeneity across an entire network. Also, the operational use of track sections is not considered in these methods. As a consequence, a new occupancy based method for the evaluation of homogeneity, including the aspect of occupancy of infrastructure, is proposed in this dissertation.

2.4 Structure of Capacity Research

The main objective of capacity research is to find out the interrelationship between the load (usually expressed as the amount of trains per unit of time) and the operation quality of the investigated system. It is an important method to evaluate the operating performance of railway operation on the infrastructure, checking whether the infrastructure is efficient utilized.

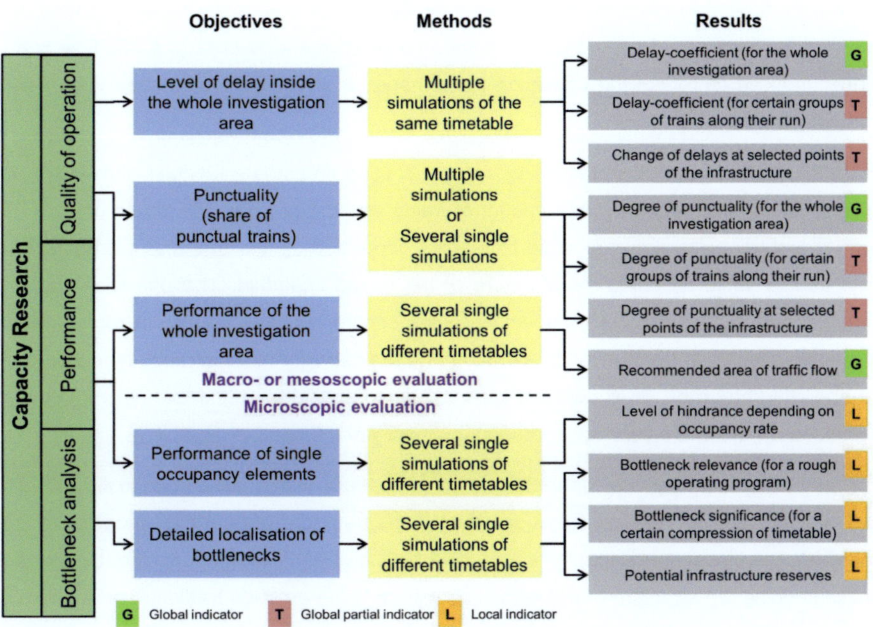

Figure 2-1:   Structure of capacity research (refer to (Martin et al. 2012))

The general structure of capacity research for railway systems is shown in Figure 2-1. The capacity research can carry out macro- or mesoscopic evaluation of the railway operation in the whole railway network, but also can evaluate the operation on single occupancy element microscopic. The level of delay and punctuality inside the whole investigated area are two major items for quality of operation. Multiple simulations of same timetables with perturbation are used to check the robustness of timetables regarding delay propagation. The degree of punctuality is significant for passengers, which is evaluated through multiple simulations. The performance of the whole investigated area is another important objective to evaluate the quality of operation and the capacity of an investigated area. Several single simulations are mainly used to determine the result of recommended area of traffic flow. For microscopic evaluation of homogeneity, the performance of single occupancy elements is evaluated through several single simulations of different timetables in order to identify the bottlenecks in the whole network.

There are two intermediate results of capacity research to determine the operating performance: the throughput capacity for an investigated area with a given operating program, and the waiting time function. Figure 2-2 is an example how to determine the waiting time function and the throughput capacity. In addition, the recommended area of traffic flow can be further induced to reflect the operating performance of the investigated area with a given operating program.

The waiting time function presents the average waiting times as a function of the load. The load is usually expressed as the amount of trains per unit of time. The single points of the waiting time diagram are stochastic timetables generated by the stepwise increase of the train density while retaining the structure of the based schedule. The operation quality is characterized by the unscheduled waiting times occurring in the system. It represents the relationship between the quality of operation and the capacity of an infrastructure with a given operating program. As shown in Figure 2-2, the capacity of the whole networks can be described by maximum capacity, throughput capacity and timetable capacity.

The maximum capacity is a theoretical value of capacity which allows congestion situation [DB Netz AG 2008]. This capacity does require the structure of operating program to keep constant. The throughput capacity is proposed by [Chu 2014], which considers the transient phase keeping the structure of operating program. It is the average traffic flow of all possible maximum throughput per time unit (one hour) at the static phase of operation process with various train sequences of a given rough operating program (train mixture). Timetable capacity is described by [Pachl 2015], which is the maximum capacity constricted to predetermined operational conditions, such as the sequence of train runs and the defined headway.

Based on the derived waiting time function with stochastic timetables, the recommended area of traffic flow can be further derived for the evaluation of operating performance [Hertel 1992]. The recommend area of traffic flow defines the traffic flow, in which traffic operates economically with satisfied operation quality. The upper and lower limit are further derived based on the waiting time function [Hertel 1992; Chu 2014]. The lower limit of the recommended area of traffic flow is given as the minimum value of the relative sensitivity function of the waiting time, which can be interpreted as the minimal change (increase) of the average waiting time from the average waiting time of the increased traffic flow. The upper limit of the recommended area of traffic flow is determined by the curve of the traffic energy function, which is the product of the number of trains (traffic flow) and the average speed per time unit.

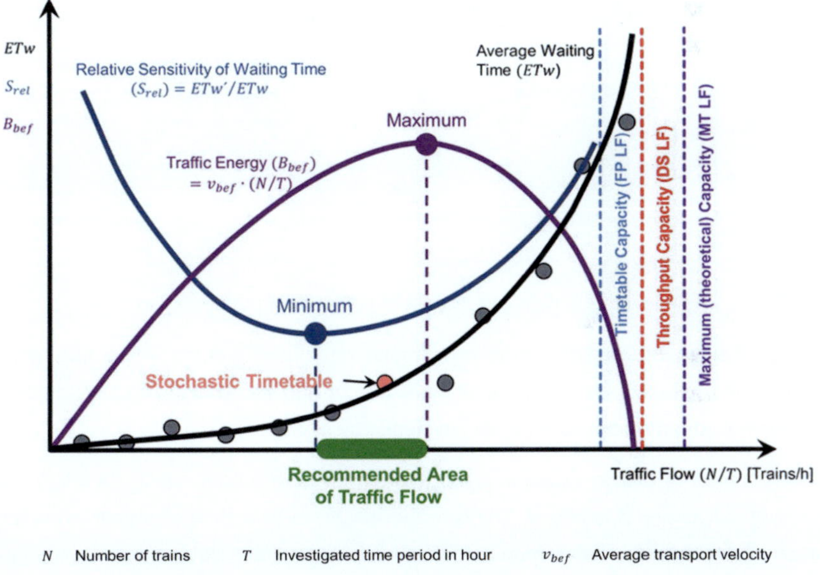

Figure 2-2:   Evaluation of performance for capacity research (refer to [Chu 2014])

It can represent the optimal relationship of capacity and quality of operation through the stochastic timetables in the investigated area within the recommended area of traffic flow. If the traffic flow is lower than the lower limit of recommended area of traffic flow, the value of waiting time is relatively small and increases moderately, which means the investigated area of railway infrastructure has lower utilization. On the contrary, if the traffic flow is higher than the upper limit of recommended area of traffic flow, the value of waiting time is high and will increase rapidly with a small growth of traffic flow.

2.5 Core Model and Software Implementations

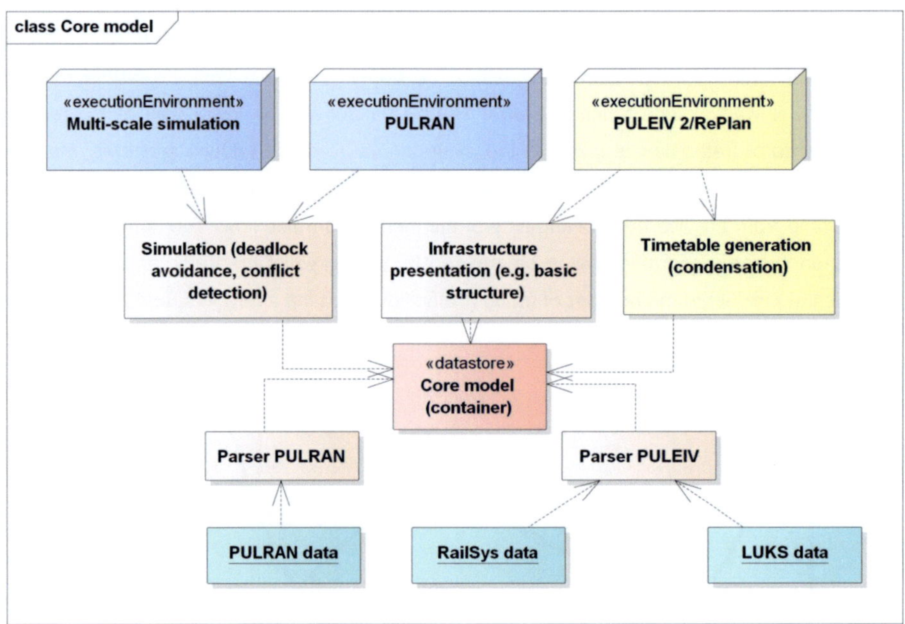

Figure 2-3:   Overview of the core model and its software implementation

In last decades, several software tools were implemented at IEV in the field of railway operations, including railway simulation, capacity research, and train dispatching. All these software tools are based on the core model developed by IEV.

The domain logic of railway systems is modelled in the core model, which covers infrastructure, rolling stocks and operations. The core model serves as a data container in the form of the internal object model. It can map data from various available data sources to the internal object model and vice versa. For this purpose, several parsers are provided for different data formats, including RailSys[12], railML[13], ISS, KSS, and other internal data formats used in IEV.

---

[12] RailSys is a synchronous microscopic simulation program for railway system, which is developed by Rail Management Consultants, Germany.

[13] RailML (Railway Markup Language) is an open, XML based data exchange format for data interoperability of railway applications.

Hence, it can integrate different software (e.g. RailSys and LUKS[14]) with the software tools developed by IEV. The core model can be conceived as a standardized layer between the external data and the individual software and application. Through isolating the details of data format of external data, developers and users can concentrate on the business logic. The implemented algorithm can be applied for different data sources based on the unified domain model.

On top of the core model, further functions and applications are built for railway simulation, capacity research and train dispatching. In Figure 2-3, an overview of the core model and the related software tools are illustrated. Some of the functions can be shared by several different software tools. For example, in order to determine the occupancy situation and the hindrances, the model of basic structures[15] is developed for capacity research [Martin et al. 2011b]. The algorithm to identify bottlenecks is based on the model of basic structures as well [Li 2015]. Furthermore, conflicts can also be identified on basic structures for train dispatching. The model of timetables is applied for railway simulation. It can also be used for generating stochastic timetables for capacity research. The same algorithm for deadlock avoidance is used for both multi-scale simulation [Liang 2017] and the investigation of shunting processes.

In this dissertation, the developed algorithm is based on the core model. Especially, the software PULEIV[16] is used as the tool for capacity research. Therefore, the developed algorithm is compatible with the existing software tools developed at IEV. Through the available parsers, it can also achieve a seamless integration with other software tools (e.g. RailSys and LUKS).

---

[14] LUKS: The LUKS software tool is an integrated system for several types of railway operations research, which is developed by VIA Consulting & Development GmbH.

[15] Basic structure: According to Martin, Li 2015, basic structure is a non-directional interrelated part of an infrastructure network. It is bounded by main signal, signal, route releasing points or the boundary of the investigated area

[16] PULEIV is a software tool developed by the Institute IEV to investigate the performance behavior of railway networks

3 Homogeneity of Operating Programs

In this section, an extended definition of homogeneity of operating programs will be introduced from the standpoint of infrastructure occupancy based on the blocking time model. This aspect is not yet considered in any existing literature regarding the homogeneity in railway operation. In addition, the existing evaluation methods of homogeneity are based on the current definitions and understanding of homogeneity, which are not suitable for the new occupancy based definition. Therefore, a new occupancy based method to evaluate the homogeneity of operating programs is also developed.

3.1 Blocking Time Model

In European railways, the blocking time model is recently widely used in capacity management and scheduling. It models the usage of infrastructure by train movements. The idea of the blocking time model was first developed by Happel in the late 1950s [Happel 1959]. In 1998, the German infrastructure operator DB Netz introduced a computer-based scheduling system based on the blocking time model. Since 2004 the blocking time theory has been recommended by the International Union of Railways [UIC 2004] for the applications in the field of railway capacity management.

3.1.1 Blocking Time

In this model, the occupancy of a track section by a train is described by blocking time. The blocking time (German: "Sperrzeit") is the total elapsed time a section of track (e.g. a block section, an interlocked route) is allocated exclusively to a train movement and therefore blocked for other trains [Pachl 2002].

In a territory with fixed block system, the blocking time of a block section for a train without a scheduled stop at the entrance of that section consists of the following time interval [Pachl 2002]:

- the time for clearing the signal;

- a certain time for the driver to view the clear aspect at the signal in rear that gives the approach indicator to the signal at the entrance of the block section;

- the approach time between the signal that provides the approach indicator and the signal at the entrance of the block section;

- the time between the block signals, as running time;

- the clearing time to clear the block section and – if required- the overlap with the full length of the train;

- the release time to "unlock" the block system

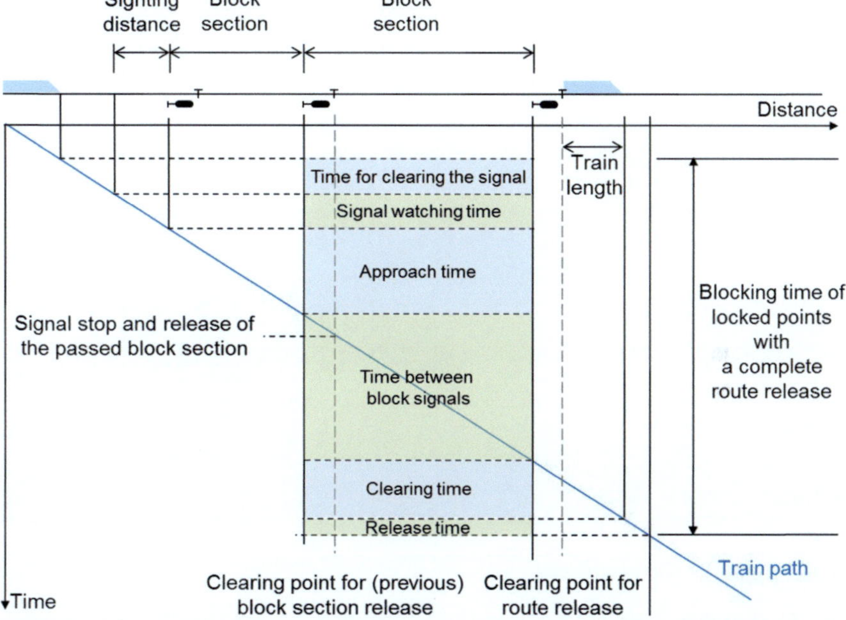

Figure 3-1:   Elements of blocking time on a block section (refer to [Pachl 2002])

As shown in Figure 3-1, the blocking time lasts from issuing a train its movement authority (e.g. by route setting) to the possibility of issuing a movement authority to another train to enter the same section. Thus, the blocking time of a track section is usually significantly longer than the time the train physically occupies the section. Therefore, comparing to running time, the blocking time is more significant variable to estimate the infrastructure occupation and to timetable scheduling.

3.1.2   Blocking time stairway

Drawing the blocking time of all block sections a train occupies along the line into a time-distance diagram17 leads to the so-called "blocking time stairway"(Figure 3-2). The blocking

---

[17] Time-distance diagram is a graph used to present the train path that consists of a time axis and a distance axis.

time stairway results from the running time calculation of a train. It soundly represents the operational use of a line by a train.

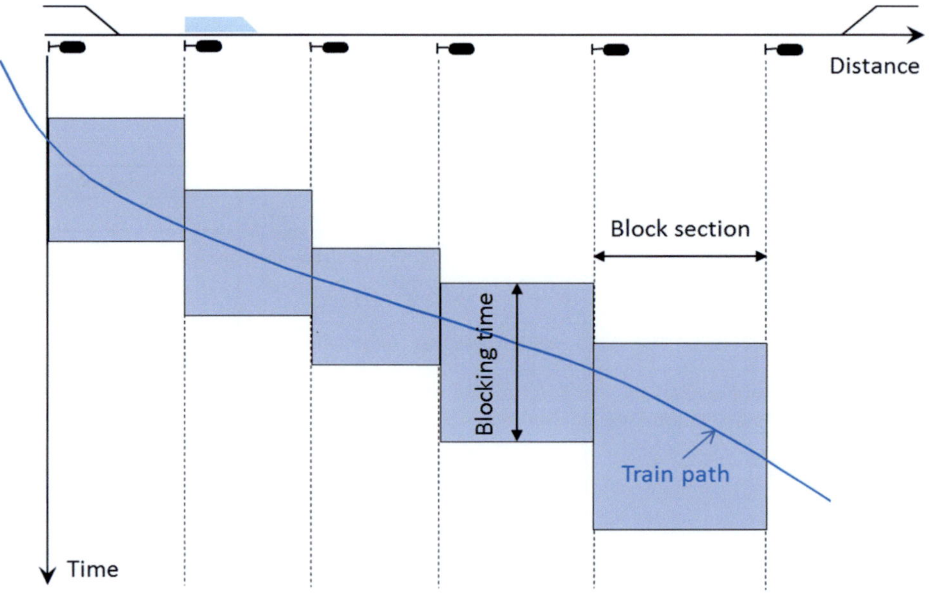

Figure 3-2: Blocking time stairway

With the blocking time stairway, it is possible to determine the minimum headway between two trains. The single headway is the minimum time interval between two successive trains considering for one block section. The minimum line headway is the minimum time lag between two trains considering not only one block section but the blocking time stairways for the entire line between two stations. In this case, the blocking time stairways of two successive trains would touch each other without any tolerance in at least one block section, which is defined as critical block section (see Figure 3-3). If there is an overlap between blocking times on a block section, a conflict can be identified.

However, during the operations, it does not matter what the absolute size of headway is, but the amount of the buffer time in this headway is. The empty time between blocking times of two trains is defined to be the buffer time, which is the time difference between the planned headways and the minimum headways. For track sections, the buffer time is the smallest slot between blocking time stairways of two trains. From each block section, buffer time (relate to train path) is the time gap (unoccupied time period) between two successive trains as well as intersecting trains. In this time interval, the block section is not occupied by any train.

A distinguish should be clear here for the recovery time and the buffer time. The recovery time is a time supplement that is added to the pure running time to enables a train to make up a small delay caused by the said train itself while the buffer time is used to prevent a small delay being transmitted from other trains to avoid so called knock-on delays.

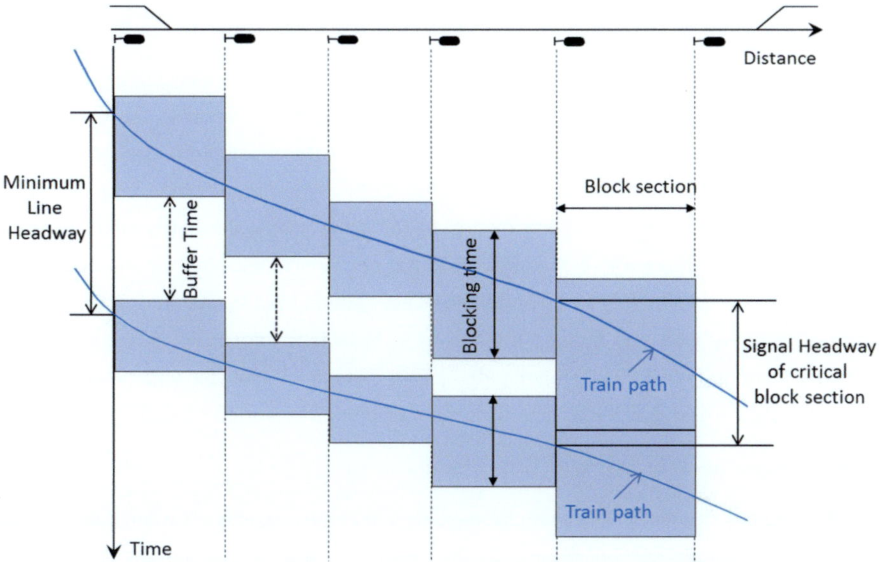

Figure 3-3:   Determining headways and buffer times by "Blocking time stairways"(refer to [Pachl 2002])

In order to reduce knock-on delays, it is relevant to find out an optimal arrangement of buffer times. In general, allocating more buffer times between trains helps the operation to recover from delays to scheduled operation more quickly. However, more buffer times have an adverse influence on the amount of trains can be operated.

The amount of buffer times depends on the required level of service, which is a trade-off between capacity and quality. Kaminsky introduced a heuristic limit for the buffer time required at bottlenecks to compensate most of the primary delays [Kaminsky 2001].

Not only the total amount of buffer times, but also their distribution in the timetable is relevant. Currently, most railway systems allocate buffer time in a deterministic way, i.e. to have fixed minimum buffer times to be added to the different kinds of train combinations. In European railway systems, a buffer is added within the range of one to three minutes. The basic rules are given as follow [Pachl 2013]:

- 1min buffer time when the second train operates at lower speed compared to the first train,

- 3min buffer time when the first train operators at lower speed compared to the second train,

- 2min buffer time when both trains operate at the same speed.

Current research activities also attempt to develop stochastic models for buffer time allocation. Kroon et al. developed a Stochastic Optimization Model to allocate the time supplements and the buffer times in a given timetable so that the timetable becomes maximally robust against stochastic disturbance [Kroon et al. 2007; Kroon et al. 2008]. Vromans developed a delay minimizing timetabling model, based on a stochastic optimization model of supplements and buffers with a predetermined set of primary delays [Vromans 2005]. With fixed total amount of recovery times and buffer times, Martin tried to find an optimization allocation of recovery time and buffer time, aiming to have less infrastructure occupation to a defined level of operation quality [Martin 2017]. But, there is no available and effective indicator to evaluate the status of buffer time allocation, whether it is a good allocation.

3.1.3 Application of Blocking Time Model

The blocking time model analysis can be used to determine capacity. The line capacity consumed by a timetable can be visualized very clearly by virtually pushing the blocking time stairways together as close as possible without any buffer time and without changing the sequence of trains. This method is also known as the "compression method" [UIC 2004]. One result of this compression process is the track occupancy, which is calculated as the ratio of the sum of the minimum line headways divided by the total time period. If the infrastructure occupation is higher than or equal to a certain typical defined value, the investigated line section shall be called congested infrastructure and no more additional train paths may be added to the timetable. If the infrastructure occupation is lower than this certain typical value, the capacity analysis must then be developed further attempting to incorporate further additional train paths into the timetable.

Beside capacity analysis, the blocking time theory is widely used in scheduling and operation [Wendler 2007]. The blocking time stairway reveals very clearly all conflicts between train paths through overlapping blocking times. Therefore, the software of advanced scheduling systems usually employs a blocking time model to support train path management. One of the characteristics of German railway operations is an entirely scheduled operation, in which the arrival and departure time of all passenger trains and freight trains are scheduled in advance. In the system, the operator works with a time-distance diagram in which a blocking time display can be switched on and off. In the case of overlapping blocking times, the operator has to solve conflicts by postponing or modifying the train paths or by changing the train sequence. Between the blocking time stairways of two trains must always be a minimum buffer time in the schedule to avoid the transmission of small delays from one train to another.

In this research, the blocking time model is the basement to identify and evaluate the homogeneity of operating programs in planning and operation processes. The form and distribution of these blocking time stairways influence the homogeneity in railway systems from the aspect of infrastructure occupation.

3.2   Variable Operating Program Structure

Existing literature reveals multiple factors influence the homogeneity of operating programs [Abril et al. 2008; Dingler et al. 2014; Hantsch et al. 2016], such as train speed, train mix, stopping pattern, etc. The variations in these factors will induce the inhomogeneity of operating programs, which can be reflected in the variations blocking time, buffer time and running direction based on the blocking time theory, which presents the occupation of infrastructure by trains. Figure 3-4**Fehler! Verweisquelle konnte nicht gefunden werden.** shows a systematic description of the influencing factors in regard to the homogeneity of operating programs.

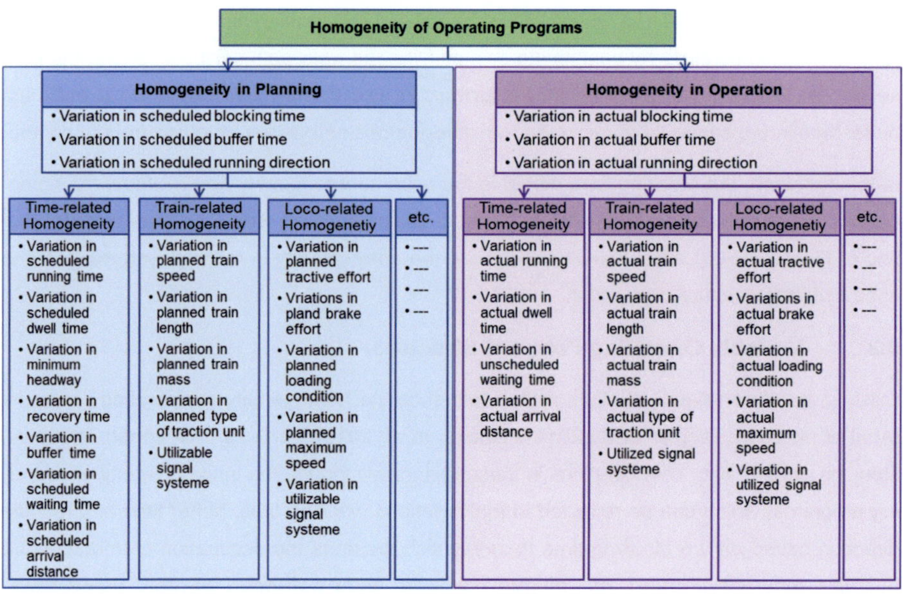

Figure 3-4:   Systematical description of the homogeneity of operating programs in railway systems (Refer to [Hantsch et al. 2016])

These factors of homogeneity can be classified into four categories: time-related, train-related, and locomotive-related, as well as the others. Locomotive-related and train-related factors describe the dynamic behavior of trains and separating of train movements. They affect the movement of each single train along the route without conflicts and scheduled stops. Besides, time-related factors include running time, dwell time, minimum headway, recovery time, scheduled waiting time and scheduled arrival distance. These factors are quite important for the robustness[18] of timetables in timetable scheduling.

In railway systems, the train speed is a very important factor for the homogeneity. It directly influences the pure running time. Besides, the accelerating and decelerating behaviors and signaling systems also affect the pure running time when the train starts and stops. The bigger the differences in train speed, acceleration and deceleration are the greater variation in running time is.

---

[18] Robustness is the ability or otherwise of a system or component to withstand model errors, parameter variations, or changes in operational conditions.

Beside the pure running time, the regular recovery time is added to every train path as a percentage of the pure running time. In European railways, the typical time supplement is 3%-7% pure running time. The scheduled dwell time is the occupancy time at schedules stops, which should be included to calculate the complete scheduled time used by the train. Scheduled waiting time is an extension of the conveyance time[19], which is introduced due to the hindrance of the investigated train paths in timetable construction. Scheduled waiting time is mostly added to the dwell time of scheduled stops but sometimes also to the running time for scheduling reasons. The variations in these influencing factors were evaluated to quantify the inhomogeneity, such as MDSR and MDFR [Landex 2008; Lindfeldt 2015].

However, this is not suitable, or exact, from the point of view of the occupancy of the infrastructure, since the operational use of a track section by a train movement is significantly longer than the scheduled running time. Besides the scheduled running time between block signals, the time for clearing the signal, signal watching time, approaching time, clearing time and release time are included in the occupancy of the infrastructure of a train, as introduce before. Therefore, running time and speed are not sufficient to present the occupation of a track by a train movement. Instead, blocking time represents the total elapsed time that an infrastructural element is allocated exclusively to a train movement and therefore blocked for other trains. The variations in train speed, locomotive related parameters, and even the train length can be reflected in the amount of blocking time.

The scheduled headways between two trains must consist of the minimum line headway plus the required buffer time to compensate for small delays. It is generally computed at stations, since calculating headway on each occupancy element is time consuming. Headways are assigned between consecutive trains according to specific combinations of trains. The operation quality will be better if the trains are evenly distributed in a time period with equal headway [Carey 1999]. Based on this analysis, changes in headways were introduced to evaluate the homogeneity, like SSHR and SAHR. However, the headway is not the blank time period between two train paths which consists of minimum line headway and buffer time. Extra time (buffer time) is added to the minimum headway between two trains to avoid transmission of small delays. Therefore, buffer time is more significant for operation quality comparing to simple headways. The amount of buffer times depends on the required level of service.

---

[19] The conveyance time results from the lining up of the running times and dwell times in the course of the train run.

Therefore, the variations in blocking time and buffer time present the homogeneous level of an operating program from the standpoint of occupancy of infrastructure. The blocking time indicates the operational use of infrastructure by a single train, while the buffer time is considered with a view to the structure of all train movements. From the aspect of occupancy, the former is the time the infrastructure is occupied by a train, and the latter is the blank of occupancy on infrastructure.

However, the variations in blocking time and in buffer time are not able to distinguish all situations. Single-track railway lines are often characterized by a very homogeneous operation with the same stop pattern, but in the case of a mix of, for example, freight and passenger trains, a more heterogeneous operation occurs. In cases of heterogeneous operation of a single track railway line, extra crossing stations may be needed due to different the running times between crossing stations varies. Especially for basic structures, the blocking time and buffer time can have same variations even with different arrangements of trains with different running directions. In Figure 3-5, the state of blocking time and buffer time is the same on the last block section, while different frequencies of direction change present different levels of homogeneity. This situation also happens on a track section or a whole network, which have an impact on the homogeneity and further on capacity obviously. Running direction is chosen as a supplementary parameter to reflect the difference in directional feature.

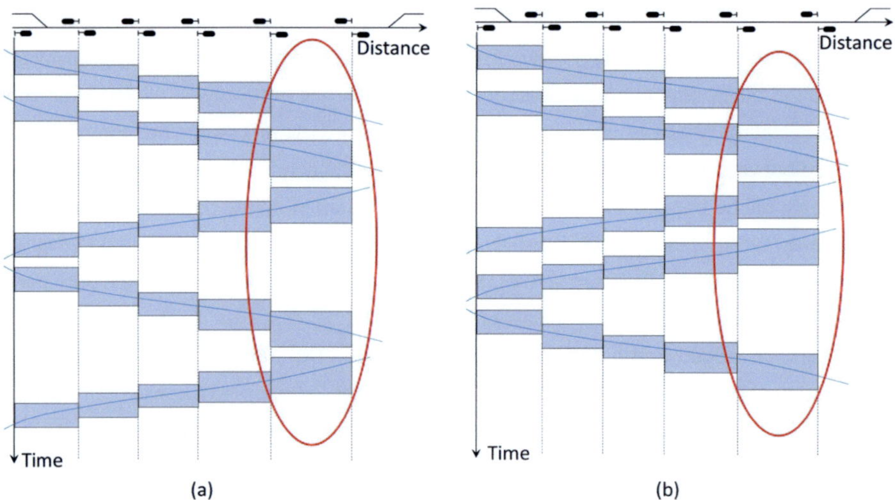

Figure 3-5:   Scenarios of different timetables related to running direction

Due to various disturbances from technical device failures, human behavior and external extreme weather conditions, the planning process of the railway operation differs from the actual operating process. Hence, the homogeneity of operating programs is considered for two scenarios: for the planning and the actual operation process within the framework of capacity research. For each process, the blocking time, the buffer time and the running direction are used to evaluate the homogeneity of operating programs. In the planning process, the homogeneity is evaluated using scheduled blocking times, scheduled buffer times and scheduled sequence of train movements. For homogeneity during the operating process, the scheduled values are replaced by corresponding values with actual or theoretical data of disturbances. In principle, during real operation, rescheduled dispatching timetables become more and more in focus.

3.3 Individual Influence from Parameters of Homogeneity

According to the system description of homogeneity of operating programs based on infrastructure occupancy, the inhomogeneity mainly originates in two aspects. First, the inhomogeneity is a consequence of different types of operated trains share the infrastructure simultaneously. These trains have different speeds, train lengths, stop patterns, dwell time, etc. The direct result of mixed train types is the variation in blocking time on occupation elements. The second aspect is the unevenly distributed train movements over given period of time having different buffer times and running direction. Therefore, conflicts will occur when not enough buffer time are arranged between consecutive trains.

3.3.1 Influence of Variation in Blocking Time

The direct consequence of the variation in blocking time is the incensement of minimum line headway between consecutive trains. For each single block section, blocking time depends not only on speed, but also signal spacing and train length. Therefore, blocking time is different for various trains and also different along the train route.

A train needs less minimum line headway comparing followed by a slower train if the consecutive train has same speed. If the train is followed by a slow train, then the headway is smaller. If the train is followed by a faster train, then time interval should be added at the beginning of the line to avoid it being caught up by fast train. Therefore, the variation influences the minimum line headway. And it increases with the variation in blocking time.

The different minimum line headways assigned between specific sequences of trains further influence the line capacity. In order to analyze the situation, essential terms are introduced here.

Occupation rate: the proportion of time occupied by train paths on an occupancy element or network elements to the whole evaluation time window. It is classified into two types according to [DB Netz AG 2008]:

- **Single occupation rate (German: Einzelbelegungsgrad)** is related to the operational use of a single occupancy element (e.g. block section, route, route section) by trains.

- **Interlinked occupation rate (German: Verketteter Belegungsgrad)** is considered for a network element (e.g. line section, line exploitation rate), which consists of several single occupancy elements. The operational feasibility should be considered for successive train runs on the network elements.

It means that minimum line headway between consecutive trains must be involved within the interlinked occupation rate. As mentioned above, the blocking time directly determines the single headways as the minimum time interval between two successive trains. In this case, the blocking time stairways of two successive trains would collide with each other without any tolerance in at least one block section.

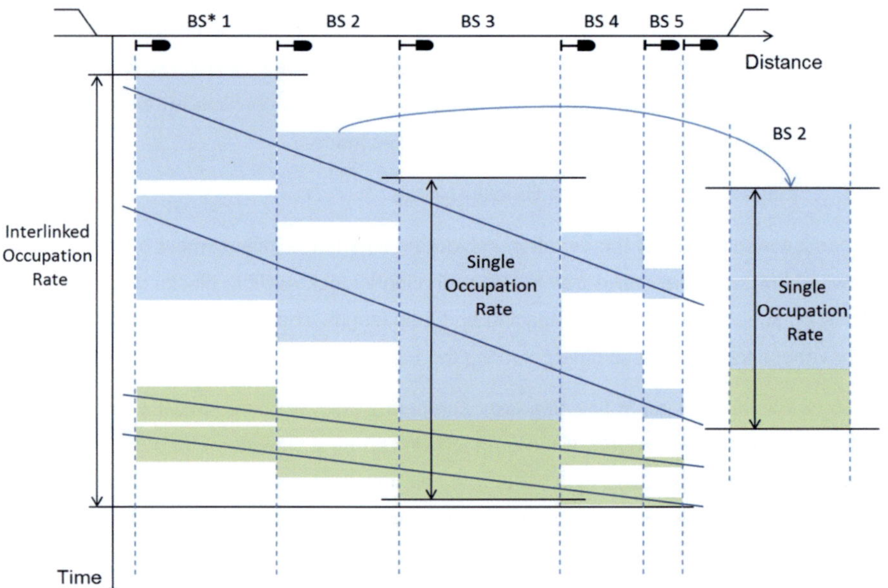

* BS= Block Section

Figure 3-6:   Single occupation rate of a block section and interlinked occupation rate of a line section

As known in Figure 3-6 interlinked occupation rate is always larger than the single occupation rates on involved block sections. There are always certain blank time periods between trains on other block sections. This blank time is set in order to eliminate conflicts between trains on critical block sections. Therefore, the interlinked occupation rate is more significant to consider the capacity of line section and networks.

Figure 3-7 illustrates how the track occupation reduces with a homogeneous operation for same traffic flow. Given the equal sum of blocking time on all involved occupancy elements, the track occupation rate is different due to the homogeneous level. It shows that through homogenizing blocking time, operation occupies the same track for less time. In practical, reducing the speed for high-speed trains is an efficient and effective measure to achieve homogeneous operation [Carey 1999]. In this case, more time can be released to avoid small delays transfer from one train to the following train. In another case, it means that new train path can be added into the network, while guarantying same operation quality.

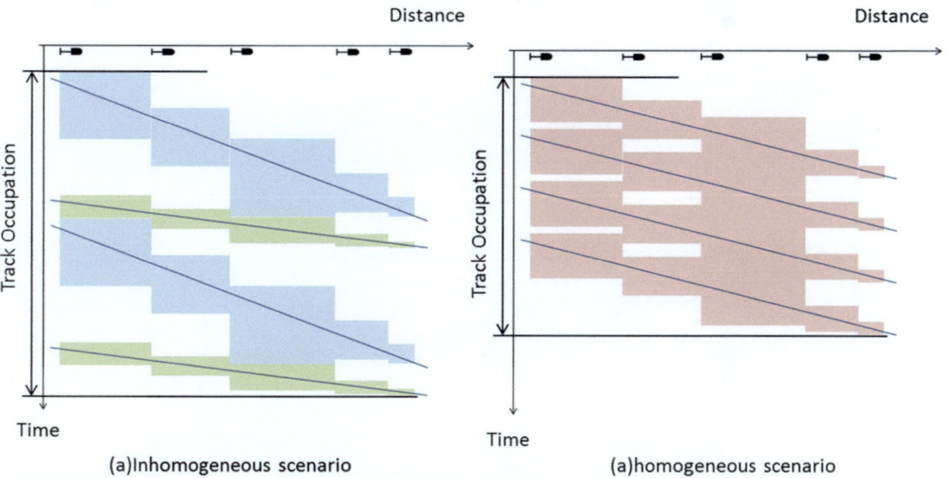

Figure 3-7:   Track occupation decreases by homogenizing blocking time

This influences the amount of trains that can be operated in the network afterward. The infrastructure occupation is calculated by compressing the timetable graphs to the limit of the critical block sections tolerance. The infrastructure occupied by a timetable can be visualized very clearly by virtually moving the blocking time stairways together as close as possible, without any buffer times but without changing the sequence of trains. This principle is known as the "compression method", which is proposed by [UIC 2004]. However, in complex timetable structures, in particular on lines with a lot of track sections, identifying the infrastructure occupation become a complicated job that requires support by a computer-based tool.

3.3.2   Influence of Variation in Buffer Time

Assuming that the first train is suffering an initial delay[20] , the delay will transfer to the second train if the buffer time is not sufficient. If the initial delay is smaller than buffer time, no knock-on delay arises in this two-train model. A high heterogeneity increases the risk for delay transfer, i.e. knock-on delays. The faster trains will catch up slower trains if not enough headways are arranged. The homogeneity of buffer time also influences the delay propagation between trains.

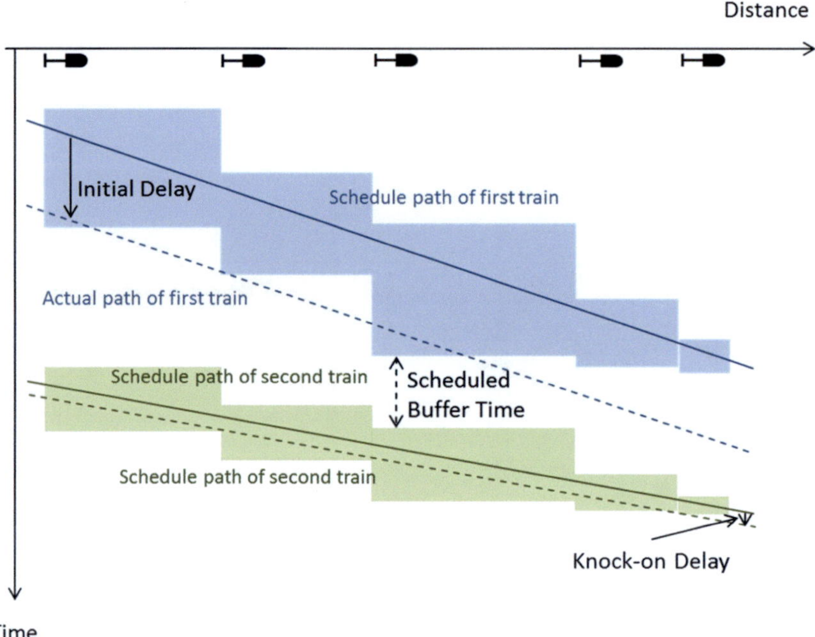

Figure 3-8:   Propagation of delay with buffer time

Usually, the amount of buffer time on mixed traffic lines depends on the kind of line headways (e.g. train speed difference) [Pachl 2002]. It is apparent that larger buffer time can re-

---

[20] Initial delay: A delay recorded at the cordons of an investigated network when the train enters that network. It is also called primary delay.

duce the possibility and magnitude of delay propagation. On the other hand, the existence of buffer time caused a reduction of operational train paths, which restricts capacity. Increasing the size of buffer time results in denser rail traffic and leads to a disadvantage in that the number of interdependencies between train routes decrease.

Assume that the total amount of buffer time available is fixed, and if 1 min more buffer time is given to one pair of train, we must take 1 min from another pair. Therefore, an appropriate allocation of buffer time between consecutive trains is significant for quality.

The sum of blocking time between trains is the same for both scenarios in Figure 3-9 Left graph (unevenly distributed train path) presents a scenario that all buffer time is allocated between last two trains, while no buffer time is available for other train combinations. The initial delay was transferred to the following train and possible further affect following trains, since no buffer time is available. In this case, even a single small delay can be transformed to next train, having greater impact on the whole network. For the scenario on right, buffer times are evenly distributed between train paths. If the delay of the first train is smaller than the buffer time, the following trains will not be affected by this delay. Even though delay of first train will transfer to the following train if the delay is larger than the buffer time, the effect on the following train will be smaller comparing to the left scenario. Even though, the sum of buffer time is the same in left graph, however, it is not utilized to avoid delay propagation. Therefore, the homogeneity of buffer time is significant of robustness of timetable face to delays and the operation quality.

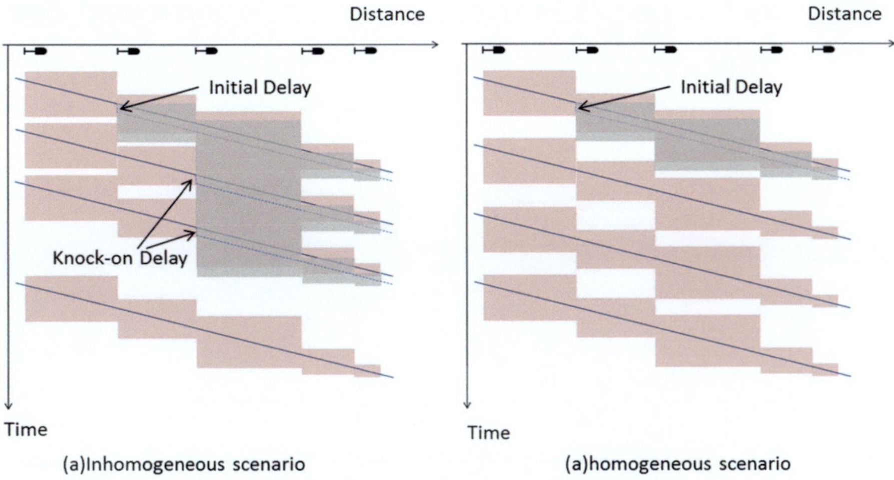

Figure 3-9:   Delay propagation decreases by homogenizing buffer time

3.3.3 Influence of Variation in Running Direction

Running direction means the direction of train movement through a track section. In railway operation, we confront the situation that trains run in different directions, causing conflicts if they attempt to use the same infrastructure elements simultaneously. This situation always happens at terminal stations, points, or on tracks that allow bidirectional running directions. However, the two parameters blocking time and buffer time are lack of directionality. When running direction changes, it means the occupation of this track section will be affected.

Take single track lines as an example (Figure 3-10). If consecutive trains have same running direction, the later can departure after the first train as long as minimum line headway is provided. The alternation of running direction requires the later train to depart after the early train completes its run on conflicting track sections where no crossing is allowed. During this time, no other train can occupy the track section. Therefore, auxiliary tracks are added to single track to enable crossing on single track.

The two scenarios in Figure 3-10 have equal amount of traffic flow, the same train mix with two running directions. Going through train movements chronologically, the train paths with the same running direction are fleeted together in the right scenario. Reversely, the trains in opposite running direction alter with each other in the left scenario. Each change of running direction constrict the fully utilization of infrastructure. The left scenario has two times more meeting of trains from opposite direction comparing to the right scenario. As a consequence, the track occupation is much longer for four trains in left scenario. The capacity is highly dependent on the "fleeting rate of direction", e.g. the frequency of changing the current direction of traffic.

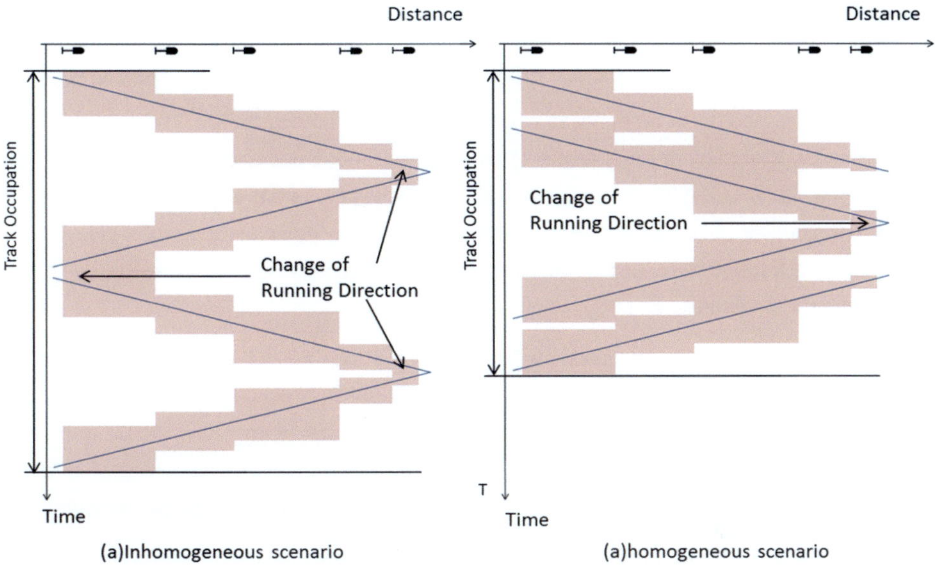

Figure 3-10:   Track occupation decrease due to homogenizing sequence on single tracks

On the other hand, fleeting of running direction leads to a larger capacity within the concerned section, it may also lead to increasing waiting times in the adjacent sections or terminals. Assembling fleets of trains may force a train that is ready to depart to wait for the next fleets of trains may force a train that is ready to depart to wait for the next fleet of its direction or speed class. The waiting time will increase with the length of the sections where fleeting is in effect. Therefore, the effects of fleeting should be carefully analyzed before implemented.

3.4   Occupancy Based Homogeneity of Operating Programs

Previous research agreed that running time differences/speed differences are key factors of an inhomogeneous operation. However, running time/speed alone cannot characterize the operational use of track sections by trains. Even trains traveling at equal speed can have different occupation times on the same track section if they have train lengths. Besides, recovery time is added to avoid delays and scheduled waiting time is added to solve conflicts during scheduling. Blocking time is relatively a more practical and significant indicator to describe the occupancy of a particular track section. The blank time interval between blocking times is defined as buffer time, which represents the temporal distribution of train running. The amount of buffer time affects the propagation of delays between trains.

If the blocking times of all trains are identical, and the buffer times between consecutive trains are uniformly spread, homogeneous railway operation is expected, such as the case of some metro systems and some high-speed rails. On the other hand, variations in the blocking time and the buffer time makes railway operations inhomogeneous. The national railway network in Germany is an example of an inhomogeneous operation that operates different types of trains with different origins and destinations, at different speeds and displaying different stopping patterns. Wherefore each of those trains demands a specific blocking time through a track section, and corresponding different buffer times are added to various sequences of train runs.

However, the variations in blocking time and in buffer time are not able to distinguish all situations. Exceptions are operating programs with equal blocking time and buffer time displaying a difference of homogeneity due to the running direction of train runs. Running direction, as a consequence, is a supplementary indicator to characterize the occupancy of infrastructure. Such situations generally occur at terminal stations, points, or on tracks allowed for opposing running directions.

In summary, **homogeneity of operating programs** (operational homogeneity) is characterized by the variation in blocking time with respect to the occupancy of track sections, variation in buffer time with respect to headways and variation in running direction with respect to the directions of train runs [Hantsch et al. 2016]. The larger these variations are, the less homogeneous the operating program is.

An operating program is defined as completely homogeneous regarding infrastructure occupancy when

- all trains have the same blocking times,

- the buffer times are entirely evenly spread, and

- there is no deviation in running directions

on the selected track sections / or in the whole network and vice versa.

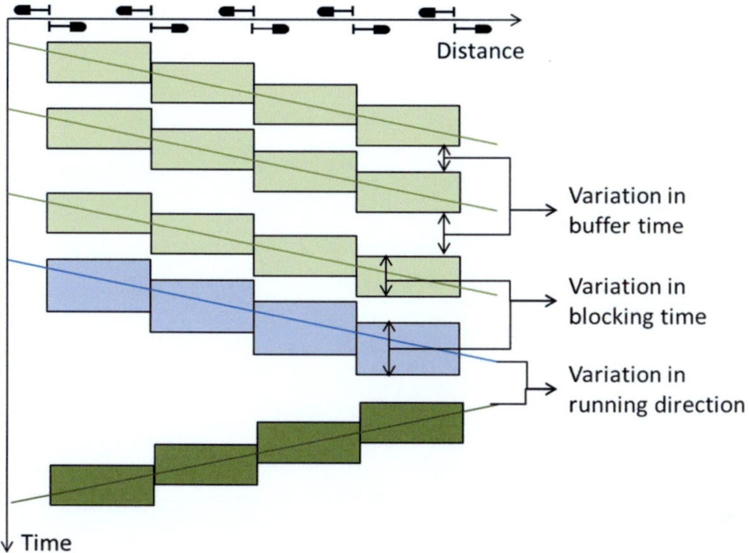

Figure 3-11:   Three parameters of homogeneity of operating programs

As shown in Figure 3-11, any variation in blocking time, buffer time and running direction makes operating program inhomogeneous. Therefore, the homogeneity of operating programs should consider all these three parameters together.

A three-dimensional model of homogeneity can be generated with variations in buffer time, in blocking time and in running direction as axis (Figure 3-11). The Euclidean Distance of individual variations in blocking time, buffer time and running direction describe the respective homogeneous magnitude of operating programs, which is inversely proportional to these variations.

This definition considers the homogeneity from the aspect of infrastructure occupation, which is also quite significant, especially for efficient utilization of existing infrastructure. It expands and complements the existing definition, having a more deeply and more comprehensively understanding of the homogeneity in railway operation. Therefore, the extended definition can be considered as a generic approach.

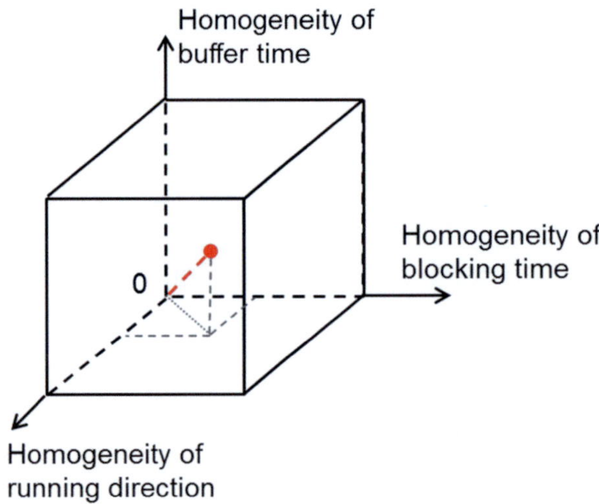

Figure 3-12:   Three-dimensional model of homogeneity of operating programs

3.5   Occupancy Based Method to Evaluate Homogeneity of Operating Programs

As mentioned above, there are multiple factors influencing the homogeneity of operating programs. Any change in these factors would lead to a change in the structure of operating program. It takes, however, an enormous amount of effort to collect the information of all these factors to evaluate homogeneity of operating programs, which is practically almost impossible. For quantitative measurement of homogeneity, the data should be easy to acquire from statistics of real operation or simulation results. As described in Section 2, existing methods use speed, running time and headway to evaluate the homogeneous level of operation. However, these methods are not suitable for the new genetic infrastructure occupancy based definition introduced in Section 3.4, in which the homogeneity of operating programs is described by variations in blocking time; buffer time and running direction in regards to the occupancy of the infrastructure. Therefore, the issue to be solved in this Section is to find the most suitable method to quantify these variations.

3.5.1 Homogeneity on Single Occupancy Elements

Homogeneity of operating programs is firstly evaluated for single occupancy elements[21]. An occupancy element is a directed or non-directed section of exclusive accessible infrastructure, whose occupation time can be determined for each train. Block section is a general example of an occupancy element, of which it is possible to obtain the blocking time. Based on a new developed microscopic infrastructure model [Martin and Li 2015; Li 2015], a railway network can be divided into several occupancy elements named basic structures (see footnote 15).

The variation in blocking time is the main reason of the homogeneity of operating programs. For each occupancy element, the operation is less homogeneous if involved trains have bigger difference in blocking times. Therefore, the coefficient of variation of all blocking times is used to describe the homogeneity regarding blocking time based on the new occupancy based definition.

Given a certain occupancy element j and an operating program, the total number of trains that occupy the occupancy element j during the observation time period is given as n. Furthermore, the start time point of the blocking time of i-th train on occupancy element j is denoted as $stBL_{ij}$ and the end time point of the blocking time of the i-th train is denoted as $etBL_{ij}$.

Then, the blocking time of the i-th train on the occupancy element j is given by:

$$tBL_{i,j} = etBL_{i,j} - stBL_{i,j} \qquad \text{Formula (10)}$$

Consequently, the expected value and the variance of blocking times on the occupancy element j are given by:

$$E_{BL_j} = \frac{1}{n}\sum_{i=1}^{n} tBL_{i,j} \qquad \text{Formula (11)}$$

$$V_{BL_j} = \frac{1}{n-1}\sum_{i=1}^{n} (tBL_{i,j} - E_{BL_j})^2 \qquad \text{Formula (12)}$$

where,

E_{BL_j} is the expected value of blocking times on the occupancy element j;

---

[21] Occupancy element (OE): An occupancy element is a directed or non-directed section of exclusively drivable infrastructure, whose occupation time can be determined for each train.

V_{BL_j} is the variance of blocking times on the occupancy element j.

With the coefficient of variation $\dfrac{\sqrt{V_{BL_j}}}{E_{BL_j}}$, the variation in blocking time of an occupancy element j can be determined through the homogeneity of blocking time (HBL_j), which is computed as follows:

$$HBL_j = \left(1 + \frac{\sqrt{V_{BL_j}}}{E_{BL_j}}\right)^{-1} \in (0, 1] \qquad \text{Formula (13)}$$

where,

$HBL_j = 1$ means "complete homogeneity in blocking time";

$HBL_j \approx 0$ means "complete inhomogeneity in blocking time".

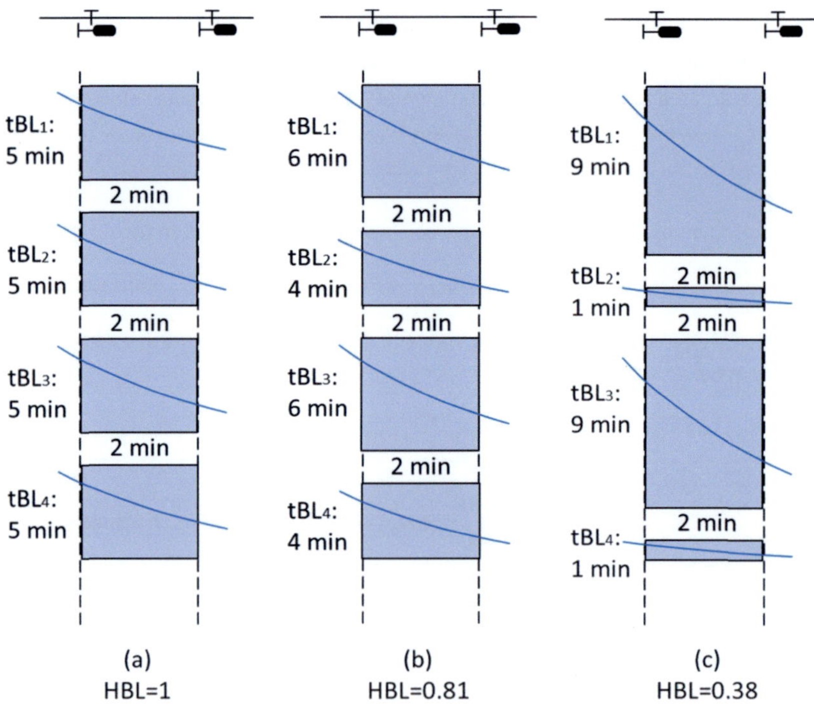

Figure 3-13:   Scenario of homogeneity of blocking time (HBL) depending on the train movements with different blocking time

Given the sum of buffer time, the increase of buffer time between a pair of consecutive trains certainly induces the decrease of buffer time between other pairs of consecutive trains. When the buffer time is not equal distributed, then the operation quality will be worsen. Therefore, the principle of evaluation for homogeneity of buffer time on an occupancy element is conducted in the same way as homogeneity of blocking time. The i-th buffer time on the occupancy element j, which is the blank time interval between blocking times of train i and $i + 1$, is given by:

$$tBU_{i,j} = stBL_{i+1,j} - etBL_{i,j} \qquad \text{Formula (14)}$$

Consequently, the expected value and the variance of the blocking times are given by:

$$E_{BU_j} = \frac{1}{n-1} \sum_{i=1}^{n-1} tBU_{i,j} \qquad \text{Formula (15)}$$

$$V_{BU_j} = \frac{1}{n-2} \sum_{i=1}^{n-1} (tBU_{i,j} - E_{BU_j})^2 \qquad \text{Formula (16)}$$

where,

E_{BU_j} is the expected value of buffer times on the occupancy element j;

V_{BU_j} is the variance of buffer times on the occupancy element j.

With the coefficient of variation $\dfrac{\sqrt{V_{BU_j}}}{E_{BU_j}}$, the homogeneity of buffer time (HBU_j), which describes the variation in buffer time on the occupancy element j can be computed as follows:

$$HBU_j = \left(1 + \frac{\sqrt{V_{BU_j}}}{E_{BU_j}}\right)^{-1} \in (0, 1] \qquad \text{Formula (17)}$$

where,

$HBU_j = 1$ means "complete homogeneity in buffer time";

$HBU_j \approx 0$ means "complete inhomogeneity in buffer time".

If buffer time evenly distributes between trains, the operation will reach an ideal homogeneous situation considering buffer time with the value of homogeneity of buffer time as 1. The value would decreases with the increase of variation in buffer times. Case (c) in Figure 3-14 is an example of an extreme inhomogeneous case, in which all available buffer time is assigned between one pair of trains, while the buffer time between other pairs of trains is set to be 0.

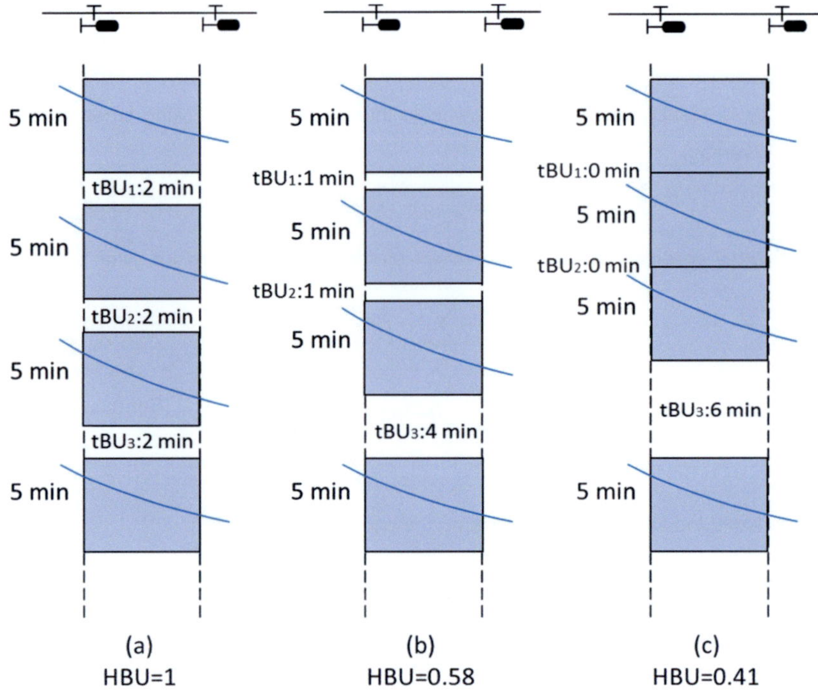

Figure 3-14: Scenario of homogeneity of buffer time (HBU) depending on the arrangement of train movements with different buffer time

Running direction means the direction of train movement through train section. For single track on a conventional double-track line, all trains run through in the same direction. In this case, the running direction of this track is homogeneous. While for a single-track line, where trains run in opposite directions, changes of running direction would happen. The railway operation is highly dependent on the "fleeting rate of direction", e.g. the frequency of changing the current direction of traffic. Each change of running direction constrict the fully utilization of infrastructure and further influence the operation quality. The more it changes, the more operation is sensitive to disturbances.

In this thesis, the change of running direction is chosen to reflect the homogeneity in regard to the aspect of running direction. As we know, on some special occupancy elements, like crossings, turn outs as well as tracks allowed for both directions, more than one running directions are allowed. When two successive trains pass through an occupancy element with different running directions, the situation is deemed that running direction changes once on this occupancy element.

Here N is given to count the changes of running direction per hour or per time period with default value as 0. Then traverse the train movements on the investigated occupancy element chronologically; add 1 to the current value of N when the running direction of one train is different from that of the forehead train. The value of N is initially set to default value of 0. The operating program is supposed to be less homogeneous with larger N regarding running directions. Accordingly, four arrangements of train movements in Figure 3-15 have the value of N equal to 5, 3, 2 and 1 separately.

As shown in Figure 3-15, the last arrangement of train movements (d) is more homogeneous than (a) (b) (c). Especially on the last block section, no variations in buffer and blocking times among trains exist, since they have opposite running directions.

The homogeneity of running direction (HRD), as the quantitative measurement of variation in running direction on the occupancy element j, can be computed with following formula:

$$HRD_j = (1 + N)^{-1} \in (0, 1]$$
Formula (18)

where,

$HRD_j = 1$ means "complete homogeneity in running direction";

$HRD_j \approx 0$ means "complete inhomogeneity in running direction".

Figure 3-15 offers four examples of the different sequences of train movements. The Case (a) is the most inhomogeneous operation considering running direction with the homogeneity of running direction as 0.17, which brings a high occupancy of infrastructure. Correspondingly, the Case (d) is the most homogenous situation among four examples.

In summary, the homogeneity of blocking time, the homogeneity of buffer time and the homogeneity of running direction represent the variations in blocking time, buffer time and running direction. In other words, they represent the overall homogeneity of operating programs based on the new definition from the aspect of infrastructure occupancy. Besides, the occupancy based method has a wide usability that can be applied to any infrastructure model, provided information of blocking times is clearly established. In addition, the characteristic of trains, such as the speed and the train length are also not required directly, since they have already been represented in the blocking time.

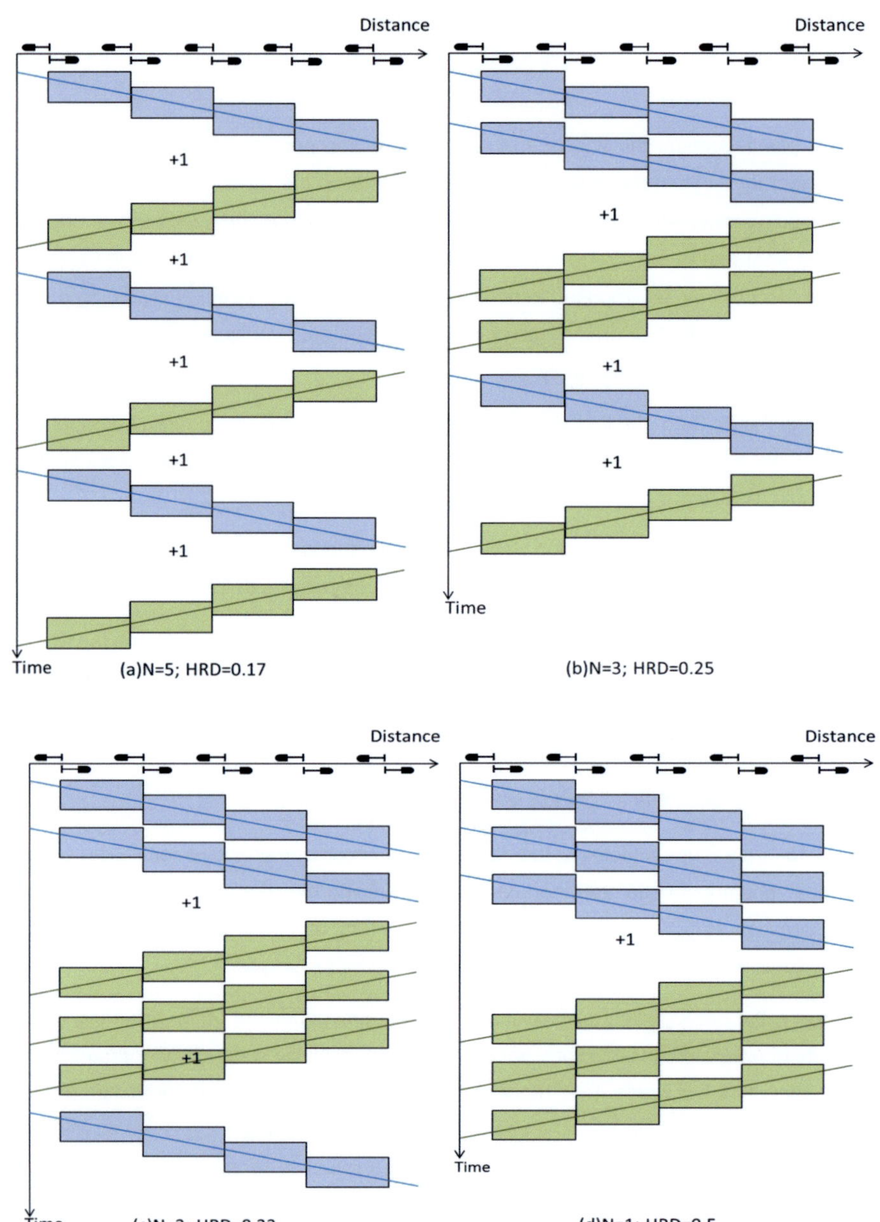

Figure 3-15: Scenarios of homogeneity of running direction (HRD) depending on the arrangement of train movements with different running directions

3.5.2 Homogeneity in an Investigated Area

The homogeneity of operating programs of a certain occupancy element is considered from the microscopic perspective. For infrastructure utilization, it is also important to get an overview of the homogeneity of operating programs in the whole investigated area, e.g. certain track sections and railway networks.

An investigated area is often a complex system. It is composed of several individual occupancy elements, whose homogeneity may have different influences for the whole network. Research showed that a clear relationship between buffer time and operation quality with high infrastructure occupation (same as low buffer rate[22] for an occupation element) is observed [UIC 2004] since litter buffer time is available to avoid delay propagation. In turn, if the infrastructure occupation is low, there is only a minor relationship between the buffer rate and delays. In other words, the homogeneity of an operating program on an occupancy element with low buffer rate plays a more important role in the operation quality of the whole investigated area. Buffer rate is the sum of buffer time on an occupancy element in regards to the blocking time stairways divided by the total duration of the investigated period, which is complementary to the occupation rate to 1 for a particular occupation element.

For homogeneous traffic, the critical block section is the block section occupied for the longest time (including time for setting up and releasing the route). This critical block section can be anywhere on the line, but often the critical block section is located close to a station or a halt due to the reduced speed when decelerating and accelerating. For heterogeneous traffic, the critical block section is usually located where the fast trains catch up with the slower trains. In the case of commuter rails, run by trains with same features, the critical section often includes high dwell time, which increases the total running time. The block section with longer blocking time is a possible position with conflicts, which is more important for the whole network. For example, long track section because of long running time and stations are significant due to the dwell time.

Therefore, buffer rate is applied to weight the influence of homogeneity of occupancy elements for the whole investigated area, that the occupancy elements with lower buffer rate are given larger weights. We assumed that m occupancy elements are involved in the investigated area. The weight for certain occupancy element j is given by:

---

[22] Buffer rate is the sum of buffer time on an occupancy element in regards to the blocking time stairways divided by the total duration of the investigated period.

$$w_j = \frac{1 - p_j}{\sum_{j=1}^{m}(1 - p_j)}$$ Formula (19)

$$\sum_{j=1}^{m} w_j = 1$$ Formula (20)

where,

m the total number of occupancy elements in an investigated area;

p_j the buffer rate on the occupancy elements j;

w_j the weight of occupancy element j.

A higher value of w_j indicates that the homogeneity of occupancy element j has more significant influence on the homogeneity of the whole investigated area.

The homogeneity of an investigated area is also calculated separately for the blocking time, buffer time and running direction. Then, the homogeneity of the blocking time within an investigated area (HBL) is computed as the weighted average of homogeneity of blocking time of individual occupancy elements:

$$HBL = \frac{1}{m}\sum_{j=1}^{m} w_j HBL_j$$ Formula (21)

where,

$HBL = 1$ means "complete homogeneity in blocking time";

$HBL \approx 0$ means "complete inhomogeneity in blocking time".

In real operation, the domain of value of HBL is distinguished for different types of networks. The commuter railways and the urban rail-bounded transportations have relatively high values of HBL over 0.9 since the trains have the same speed. But in a mixed traffic network, HBL normally ranges from 0.6 to 0.9. The value of HBL below 0.6 rarely happens in the real railway operation.

The homogeneity of buffer time in an investigated area (HBU) is computed as the weighted average of homogeneity of buffer time of individual occupancy elements:

$$HBU = \frac{1}{m}\sum_{j=1}^{m} w_j HBU_j$$ Formula (22)

where,

$HBU = 1$ means "complete homogeneity in buffer times";

$HBU \approx 0$ means "complete inhomogeneity in buffer times".

Normally the value of HBU has to be assumed in the range between 0.6 and 0.8 in real railway operation. Even in the commuter railways and the urban rail-bounded transportations, the HBU is restricted due to uneven distribution of train paths arrangement in the whole network.

It is the same for the calculation of the homogeneity of running direction within the investigated area (HRD) as the weighted average of homogeneity of running direction of individual occupancy elements:

$$HRD = \frac{1}{m} \sum\nolimits_{j=1}^{m} w_j HRD_j \qquad \text{Formula (23)}$$

where,

$HRD = 1$ means "complete homogeneity in running direction";

$HRD \approx 0$ means "complete inhomogeneity in running direction".

HRD is always relatively small below 0.5 on single track lines. But on conventional double track lines, the trains from different directions run separately on their respective tracks, giving the value of HRD as 1. For the whole network, the change of running direction always happens at crossings, stations or on single tracks, which are just a part of the whole network. Therefore, the value of HRD in a whole network is always between 0.6 and 0.8 in real operation.

The sketch of a small complete infrastructure network of a reference example is presented in the Figure 3-16. The network consists of 4 stations (AHX, BS, LBC, and EN) and 68 occupancy elements in total. The stations EN and BS are connected by a single-track line, and the others are double-track lines.

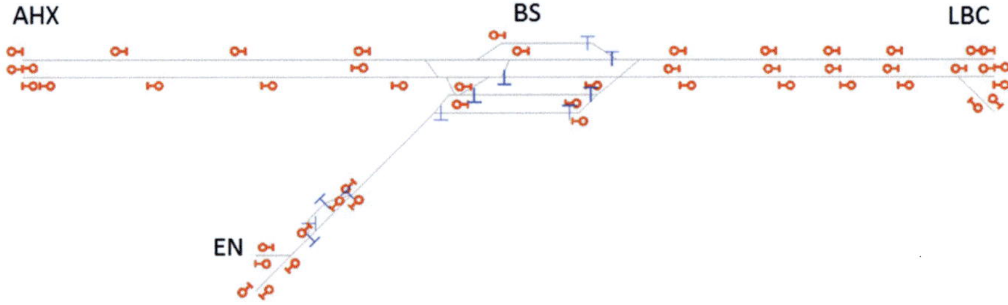

Figure 3-16:   the infrastructure network of the investigated area

The homogeneity of operating programs was evaluated in the whole network for three operating programs scenarios (shown in Table 3-1) through the method described above. Three types of trains are involved: long distance passenger trains (FRZ), short distance passenger (NRZ) and freight trains (GV). Four running directions are defined: from station AHX to station LBC, from station LBC to station AHX, from station EN to station LBC and from station LBC to station EN.

If the same type of trains operates in the network, like scenario (c), the operation will reach a comparatively homogeneous situation with the value of HBL as 0.95, HUB as 0.69 and HRD as 0.77. Each value would decrease with variations of train types. Case (a) in **Fehler! Verweisquelle konnte nicht gefunden werden.** is an example of a comparatively inhomogeneous case, in which three different types of trains share the network simultaneously. In Case (a), the value of three parameters of homogeneity respectively are 0.83(HBL), 0.67(HBU), and 0.69 (HRD), which are smaller than the corresponding value in scenario (c).

Table 3-1:   Scenarios of homogeneity of operating programs in the network example

(a)   **(b)**   **(c)**

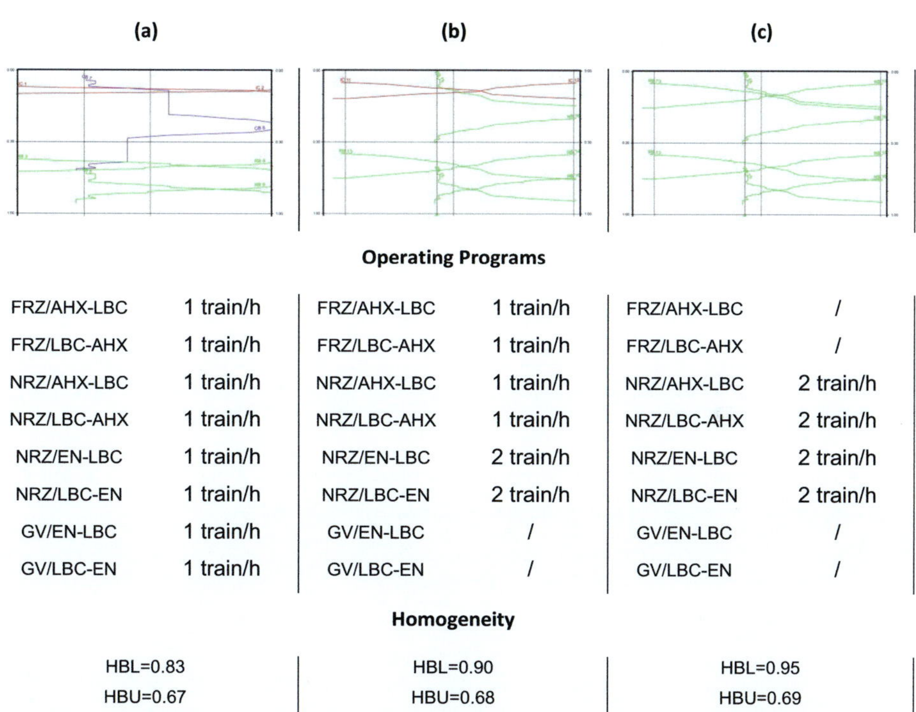

Operating Programs

(a)		(b)		(c)	
FRZ/AHX-LBC	1 train/h	FRZ/AHX-LBC	1 train/h	FRZ/AHX-LBC	/
FRZ/LBC-AHX	1 train/h	FRZ/LBC-AHX	1 train/h	FRZ/LBC-AHX	/
NRZ/AHX-LBC	1 train/h	NRZ/AHX-LBC	1 train/h	NRZ/AHX-LBC	2 train/h
NRZ/LBC-AHX	1 train/h	NRZ/LBC-AHX	1 train/h	NRZ/LBC-AHX	2 train/h
NRZ/EN-LBC	1 train/h	NRZ/EN-LBC	2 train/h	NRZ/EN-LBC	2 train/h
NRZ/LBC-EN	1 train/h	NRZ/LBC-EN	2 train/h	NRZ/LBC-EN	2 train/h
GV/EN-LBC	1 train/h	GV/EN-LBC	/	GV/EN-LBC	/
GV/LBC-EN	1 train/h	GV/LBC-EN	/	GV/LBC-EN	/

Homogeneity

(a)	(b)	(c)
HBL=0.83	HBL=0.90	HBL=0.95
HBU=0.67	HBU=0.68	HBU=0.69
HRD=0.69	HRD=0.74	HRD=0.77

* The train paths of FRZ are marked in red, the train paths of NRZ are marked in green and the train paths of GV are marked in violet.

3.6   Comparison of Occupancy Based Method and Existing Methods

A comparison is conducted between existing methods and the new occupancy based method for evaluating the homogeneity of operating programs in this section. As described in Section 2.3.2, several approaches have been developed in previous research. In this dissertation, the precision and effectiveness of the occupancy based methods is compared to three main methods, which are developed by Vromans [Vromans 2005], Landex [Landex 2008] and Lindfeldt [Lindfeldt 2015].

- Vromans created indicators SSHR and SAHR to evaluate homogeneity of operation. If the running time of each train and the order of trains running are pre-specified, the operation reaches a homogeneous situation when the headways between subsequent trains are equalized during the whole time spread. However, these two indicators are not comparable if the sum of minimum headways is different.

- Landex calculated the ratio of SAHR and SSHR as an indicator, whose value is 1 in homogeneous operation and 0 in inhomogeneous operation. Nevertheless, this ratio is still equal to 1 if the trains are operated with different headway times, which is not homogenous operation. Therefore, the other ratio of the headway at departure station to the following headway multiplied by the ratio of headways for arrival at stations is further improved. However, this ratio is mathematically more complex to calculate.

- Indicators MDSR and MDFR were presented by Lindfeldt as supplements to indicators SSHR and SAHR in order to describe the differences in running time based on the timetable. A homogeneous operation has the value of MDSR and MDFR as 0.

Three existing methods and the new occupancy based method are applied to evaluate the homogeneity in the following two examples as shown in Figure 3-17.

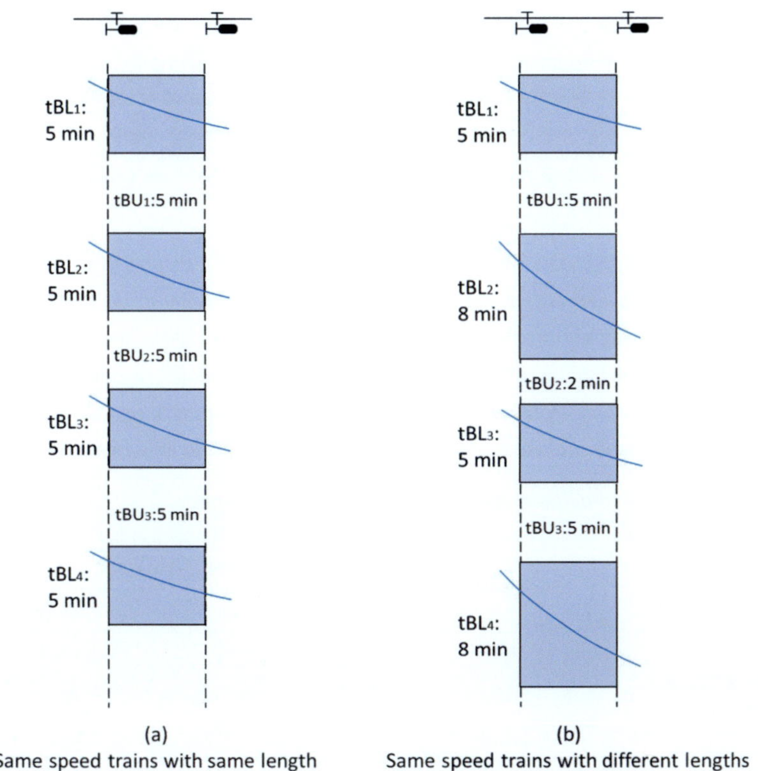

Same speed trains with same length   Same speed trains with different lengths

Figure 3-17:   Scenarios for comparing different evaluation methods of homogeneity

Scenario (a) is an absolute homogenous situation that identical trains with same speed and train length run through an occupancy element with equal headways between them. Apparently, the trains demand identical length of running time and blocking time to through the occupancy element. Meanwhile, they are evenly distributed on the infrastructure element with same headways and buffer time.

In scenario (b), the blocking times on the occupancy element are different to trains due to various train lengths, even though the trains operate at equal speed. We assume that the headways between trains remain same as scenario (a); these trains consequently have different buffer times between them.

The values of homogeneity are calculated for these two scenarios by four methods and the results are presented in the following Table 3-2. For scenario (a), Vromans's method gives the value of SSHR as 0.30 and SARH as 0.3. However, this scenario cannot be defined as homogeneous only based on this value. By other three methods, the values of homogeneity indicate the absolute homogeneous state of scenario (a). Comparing scenario (a), the results of Vromans's, Landex's and Lindfeldt's methods in scenario (b) are same. It means scenario (b) is also homogeneous. But, from the perspective of infrastructure occupancy, the operation is not homogeneous. These variations in blocking time and buffer time can be reflected by indicators HBL, HBU and HRD.

In conclusion, all methods discussed here are capable to quantize the degree of homogeneity of operation in railway systems to some extent. They can identical the absolute homogeneous operation for same speed trains with same length. But the existing methods cannot recognize the inhomogeneity due to differences of train length, which influences the length of blocking time. The new occupancy based method, developed in this thesis, can also quantify these differences in the parameters of homogeneity, which is a tribute to its advantage.

Table 3-2:   Comparison of different methods to evaluate homgoeneity

Method	Indicators	Homogeneity Value	
		Scenario (a)	Scenario (b)
Occupancy Based Method	HBL	1	0.79
	HBU	1	0,70
	HRD	1	1
Vromans's Method	SSHR	0.30	0.30
	SAHR	0.30	0.30
Landex's Method	SAHR/SSHR	1	1
	Homogeneity Rate	1	1
Lindfeldt's Method	MDSR	0	0
	MDFR	0	0

4 Influence of Parameters of Homogeneity

Based on the occupancy based definition of homogeneity of operating programs, the homogeneity can be characterized by variations in blocking times, buffer times, and running directions. These variations can subsequently be evaluated respectively through parameters, e.g. the homogeneity of blocking time (HBL), the homogeneity of buffer time (HBU) and the homogeneity of running direction (HRD) as described above in Section 3.4 and 3.5. It also reveals that these parameters can well reflecting the change in operating programs regarding blocking time, buffer time and running direction. In addition, the method can evaluate the homogeneity not only for a single occupation element, but for an entire network. In this Section, the influence of homogeneity of operating programs on the operation quality is investigated separately for each parameter of homogeneity (HBL, HBU and HRD). A case study on a real commuter railway network was conducted, followed by the validation of influence of each parameter.

4.1 Investigated Area

The analysis is conducted on a real commuter rail network in Germany (see Figure 4-1), in which it is adapted to operate with both absolute homogeneous and inhomogeneous operating programs.

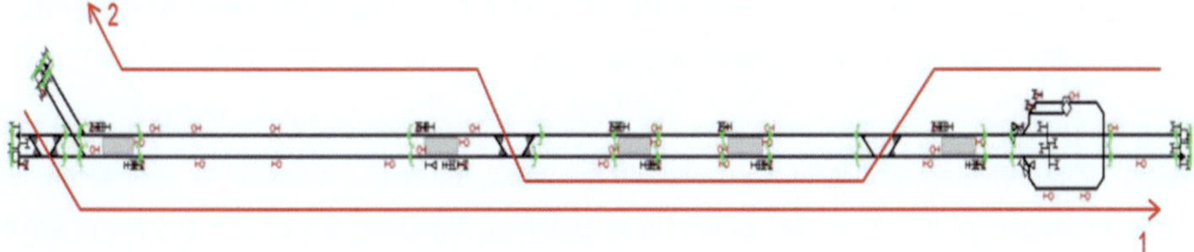

Figure 4-1:   Layout of a Commuter Rail Network (S-Bahn[23])

This network consists of two double track lines with 9 stations in total. Two routes of opposite directions share the same track section in the middle on which running direction can be changed when needed. In real operation, the trains of opposite directions operate separately on their own tracks, leaving no conflict with the trains from the opposite direction. Nevertheless, in order to analyze the influence coming from homogeneity of running direction, some

---

[23] S-Bahn is a type of hybrid urban and suburban railway. Some of the larger S-Bahn systems provide services similar to rapid transit systems, while smaller ones often resemble commuter or even regional rail.

adjustments have been made to Route 2 to create some conflicts as shown in Figure 4-1. Route 1 is composite of 14 occupation elements and Route 2 has 11 occupation elements in total.

In this network, only one type of train is operated in the network with the headway as 5min. All these trains are the same type with the same maximal speed 140km/h and identical accelerating and decelerating behaviors. The only difference among them is the train length. The different train lengths make the blocking times on each occupation element vary from one another. The investigated time period is from 0:00 to 6:00.

4.2 Research Tools

Two types of software are utilized in the analysis: RailSys simulates train movement in real operation; PULEIV evaluates the capacity performance.

- RailSys

RailSys is a synchronous microscopic simulation program for railway systems, which is developed by the Rail Management Consultants, Germany. It consists of the four program elements: Infrastructure Manager, Timetable Manager, Simulation Manager and Evaluation Manager, which has been successfully applied in plenty of projects [Aly et al. 2016; Martin 2017]. In this study, two functions were applied: Timetable Manager to generate timetables possessing varying levels of heterogeneity and Simulation Manager to simulate operation of those artificial timetables from Timetable Manager.

Inhomogeneous timetables are generated based on an absolute homogeneous timetable with Timetable Manager. With the timetable manager, not only the timetable parameters but also properties of trains can be defined and modified. Modifications on the original absolute homogeneous timetable, e.g. altering train length of related trains, modifying dwell time at associated stations, shifting departure time of relevant trains and changing the sequence of train runs, makes the original absolute homogeneous timetable inhomogeneous

When timetables are generated, timetable simulation should be conducted for each timetable by the Simulation Manager of RailSys. It simulates the real operation process of a given timetable. Through the timetable simulation, the time point of each train through measurement points will be recorded in a protocol. This data would be utilized later to calculate the parameters of homogeneity by using the software tool PULEIV. A short description of PULEIV will be given in the following paragraph.

An operational simulation would also be conducted to evaluate the operation quality. The input disturbance is expected to be negative exponential distributed. The disturbance parameters used to generate input disturbances in the operational simulation are shown in the following table:

Table 4-1:   Disturbance parameters of input disturbance (refer to [DB Netz AG 2008])

Disturbance Parameters	Initial Delay	Dwell Time Extension
Probability of Delay [%]	25	5
Average Delay [min]	2	0.2
Maximum Delay [min]	15	5

- PULEIV

PULEIV is a software tool developed by the Institute IEV to investigate the performance behavior of railway networks, which contains several functions, in particular for capacity research. For example, PULEIV can determine the recommended area of traffic flow based on waiting time function. It can also identify and locate bottleneck and evaluate the operation quality. So far, PULEIV was successfully applied in several projects [Martin et al. 2008; Martin et al. 2013; Martin et al. 2011a].

PULEIV is utilized, in this research, to calculate the three parameters of homogeneity of operating programs. This software can divide the whole network into single occupancy elements (e.g. basic structures) and extract the start time and end time for each train on each single occupancy element from the simulation protocol of timetable simulation. The calculation process of parameters of homogeneity of operating programs can be further implemented according to the method described in Section 3. Besides, the operation quality could be evaluated through PULEIV by the indicator delay-coefficient based on the simulation protocol of operational simulation. More specifically, the delay-coefficient is calculated as the ratio of out-coming delays and in-coming delays through the whole network. When the value of delay-coefficient is less than 1, it indicates that the timetable is robust, which is capable of eliminating some delays in this system. The change in operation quality, due to the heterogeneous operating program, would be represented in the difference of delay-coefficient observed between original homogeneous timetable and created inhomogeneous timetables.

4.3   Case Study

The structure of an operating program is significant for railway operation. When the traffic flow[24] is constant, a better operation performance will be achieved with a more homogeneous operating program, like higher punctuality and smaller average waiting time [Abril et al. 2008; Lai and Barkan 2011; Pachl 2013]. The influence of parameters of homogeneity on operation quality would be quantitatively investigated in this section for certain traffic flow.

Figure 4-2 describes the workflow for completing the influence analysis of homogeneous paramters. The analysis starts with an absolute homogeneous timetables, in which identical trains at same speed even spread in the investigated time period. The basic princple is to evalute the operation quality of timetables with different levels of homogeneity and then figure out the interrelationship between operation quality and corresponding parameter of homogeneity. Therefore, the analysis mainly contains two important components.

Firstly, a set of timetables posessing different values of parameters of homogeneity is required. For different parameters, the ways to generate these timetables are different. With a view to the components of blocking time as introduced in Section 3.1, the changes in light of running time, dwell time, clearing time, etc. brings about an alternation in blocking time. For example, longer trains need more time to run through the clearing points of each block section. The blocking time in a station area always coincides with the change of the scheduled dwell time.

Buffer time is the time gap between the blocking time of train pathes. The amount of buffer time is liable to be influenced by shifting the train path in the investigated time period, thus the variation in buffer times will be induced. When the order of train runs is different, the homogeneity of running time will accordingly make a difference in value of running direction.

The second important task is, afterward, to assess the operation quality of these various timetables artifically generated by the methods introduced above. In this analysis, the indicator dealy-coefficient[25] is used to present the operation quality of each timetable with certain disturbances. As introduced above, this indicator can be calculated through software PULEIV to calculate the parameters of homogeneity of each artifical generated timetable based on the potocal of timetable simulation. The change in operation quality, due to the

---

[24] Traffic flow: Traffic flow is the number of train runs as throughput in a time unit during the investigated time period (refer to (DB Netz AG 2008)).

[25] Delay-coefficient is the ratio of out-coming delays and in-coming delays through the whole network.

heterogeneous operating program, would be represented in the difference of delay-coefficient observed between homogeneous and developed inhomogeneous timetables.

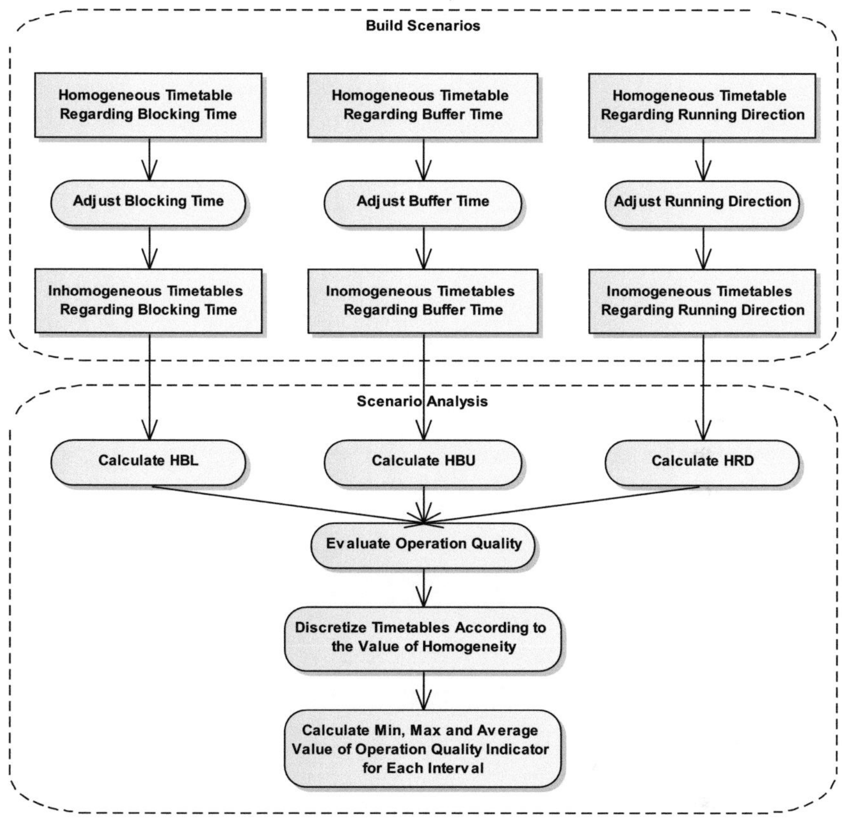

Figure 4-2:   Workflow for the analysis of parameters of homogeneity

The operation quality of these timetables with different level of homogeneity are interested, whether they are changed due to the homogeneous level. The correlationship between operation quality and each parameter of homogeneity was studied by scattor plot. As the Section 3.5 illustrated, the value are continous data from 0 (inhomogeneous) to 1 (homogeneous). The timetables were broken up based on the value of homogeneity in an effect to establish equal discretization into 20 areas with interval as 0.5. The minmum, maximum and average value of the operation quality evaluated in a homogeneity interval were calculated.

4.3.1   Influence of Homogeneity of Blocking Time

For the analysis of individual influence from the homogeneity of blocking time, only the Route 1 is of interest (see Figure 4-1), since the situation of the opposite direction is the same. In the original homogeneous timetable, there are 69 identical trains run by Route 1 with headway 5 min, complying with real operation during high peak hours. The first train departs at 0:01:37 and last train at 5:41:37. To ensure that the total operational time is the same for all timetables, the first train and the last train won't make any changes, including departure time, train length or dwell times.

First is how to make a change on the blocking time, and further, induce the variation in blocking time. In this study, two measures to alter the homogeneity of blocking time were adopted. One is altering the train length; the other is changing the dwell time. Examples of timetables generated through different ways are presented in Table 4-2.

Table 4-2:   Scenarios of heterogeneous timetables regarding blocking time

Timetable	Train Group ***	Train Length [m]	Blocking Time* [s]	Dwell Time** [s]	Homogeneity of Operating Programs		
					HBL[-]	HBU[-]	HRD[-]
Case 1	S1	136	1304	174			
	S2	136	1304	174	1	1	1
	S3	136	1304	174			
Case 2	S1	1	1130	174			
	S2	136	1304	174	0.97	0.69	1
	S3	207	1510	174			
Case 3	S1	1	1130	246			
	S2	136	1304	174	0.95	0.64	1
	S3	207	1510	102			
Case 4	one train	136	1800	680	0.63	0.54	1
	other trains	1	1510	174			

*: the sum of blocking time through each occupation element for a train;

**: the sum of dwell time at each station for a train.

***: the first train is excluded in train group S1, and the last train is excluded in train group S3

Increasing train length would lead to longer release time for the train to leave an occupation element. On the contrary, the blocking time will decrease with shorter trains. The extending and shorting of the train length will always be done concurrently to ensure the total blocking time is almost the same for all timetables. However, the magnitude of change is limited by this method because the difference in blocking time between the longest and shortest train is not dramatic. It is either not reasonable to infinitely extend or shorten train lengths in actual process.

The other way to change the blocking time is to extend the dwell time at stops. In practice, a train may wait at a station or in front of a signal to avoid conflicts or ensure connections, especially for freight trains. The change of homogeneity of blocking time through this method is also constricted due to the length of the investigated time period. In some timetables, both actions are therefore combined to generate a larger variation in blocking time, like Case 3 and Case 4.

Table 4-2 also suggests that the value of homogeneity of buffer time is also changed. It is because; alterations of the blocking times are accompanied by changes in the buffer times on single occupancy elements since the operating time is constant for all timetables. In addition, the degrees of these changes may be different along occupation elements of a train path due to the length of occupation elements and some limitation of infrastructure, e.g. speed limit. In some cases, even the critical occupancy element between a pair of trains was changed. Due to different magnitudes of changes in buffer time on single occupancy elements, it is not possible to make buffer times evenly distributed on all occupancy elements. Therefore, the train paths were adjusted to spread evenly on critical block sections to minimize the influence from unevenly distributed buffer times on operation quality.

The value of homogeneity of blocking time (HBL) is always over 0.4, which is because of the limitations of train length and restrictive investigated period. Train length only influences the clearing time through the release point, while clearing time is actually a small part of the blocking time. Besides, the train length and the dwell time cannot be extended without restrictions, and even run through the whole investigated period. In real operation, the train mix should comply with the various traffic demands. Nevertheless, this train mix determines the domain of values of HBL.

The scatter plot (Figure 4-3) visually demonstrates the increase of delay-coefficient over the homogeneity of blocking time. In general, the delay-coefficient tends to increase along with the decrease in the homogeneity of blocking time. The change of delay-coefficient is relatively small especially for timetables with a high value of homogeneity of blocking time over 0.9. While, along with the continuous decrease of HBL, the increase of delay-coefficient is greater accordingly.

Another argument is that the delay-coefficient fluctuates for a certain value of HBL. Even the homogeneity of blocking time is not so good; it can also reach to a good operation quality with a suitable arrangement of train runs. Therefore, the delay-coefficients of various inhomogeneous timetables are averaged for each interval. The line illustrated the positive relationship between delay-coefficient and homogeneity of blocking time.

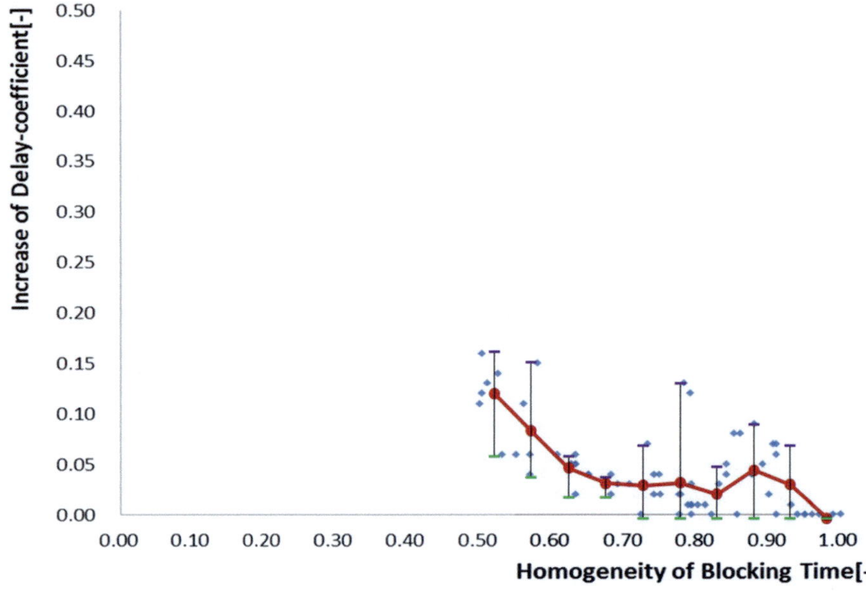

Figure 4-3:   The relationship between homogeneity of blocking time and delay-coefficient

Investigating the reason, the variance in infrastructure occupation corresponding to the homogeneity of blocking time was interested. With the decrease of homogeneity of blocking time, the infrastructure occupation has an increasing tendency. The operation quality trends consistent with the change of infrastructure occupation. The influence of homogeneity of blocking time on operation quality is accordingly in consequence of infrastructure occupation. The increase of infrastructure occupation would result in less buffer time can be used to deal with delays. When a slow train follows a fast train on an occupation element, the time gap between them grows over the route. At the end of the route, the amount even excesses the required amount to compensate delays. This valuable time cannot be used by other trains on this occupation element, which has a negative influence on the operation quality. The greater the variation in blocking times, the more time gaps between train paths will be available. These changes subsequently influence the operation quality. In other words, the influence on operation quality from the homogeneity of blocking time is through the increase of infrastructure occupation, which reduces the available time interval for new trains or buffer time.

4.3.2   Influence of Homogeneity of Buffer Time

The influence analysis of homogeneity of buffer time on operation quality is based on the same complete homogeneous timetable on Route 1, as in the analysis of homogeneity of blocking time. There are 69 trains run through the route at the same speed in total. Same as the analysis in Section 4.3.1, the first train and the last train won't make any changes to keep the usable operation time constant.

Table 4-3:   Scenarios of heterogeneous timetables through shifting train path

Timetable	Train Combination	Buffer Time* [s]	Headway** [s]	Homogeneity of Operating Programs		
				HBL[-]	HBU[-]	HRD[-]
	S1-S2	152	300			
Case 1	S2-S3	152	300	1	1	1
	S3-S1	152	300			
	S1-S2	132	280			
Case 2	S2-S3	172	320	1	0.96	1
	S3-S1	152	300			
	S1-S2	92	240			
Case 3	S2-S3	212	360	1	0.75	1
	S3-S2	152	300			
Case 4	First 67 Pairs	0	148	1	0.54	1
	Last Pair	10336	20252			

*: Buffer time on critical block section;

**: Headways at the start of train path.

In this section, inhomogeneous timetables regarding buffer time are generated through shifting train paths inside the investigated time period. Shifting a train path, through changing the departure time, will alter the buffer times of this train between the previous and following trains, further altering the temporal distribution of buffer time. The buffer time can reach minimum as 0 min between consecutive trains, guaranteeing no conflicts. Table 4-3 gives 4 examples of different scenarios of buffer time distributed in the investigated time period. The amount of headways between train paths were also changed in the process of shifting train path. Well, this method can dramatically change the value of homogeneity of buffer time spreading over the whole range. This value approaches 0.54 in Case 4, in which ahead 68 trains are compressed without any buffer time leaving a huge time gap between the last two trains. This is a virtual extreme situation for the analysis. In the real railway operation, however, this value is subjected to the requirement of minimum buffer time between different train combinations, as discussed in Section 3.1.

The relationship between homogeneity of buffer time and operation quality is depicted in the following scatter plot. The scatters are distributed in the whole range of the homogeneous value. In comparison to the interrelationship between homogeneity of blocking time and operation quality, the figure indicates a more remarkable trend; that the delay-coefficient increases as the homogeneity of buffer time decreases. As it can be observed, the delay-coefficient on this route is stable relatively low for the growth of delay-coefficient which corresponds to the value of homogeneity of buffer time over 0.8. The timetables with the HBU in this range always have good ability to deal with the primary delays. For the value of homogeneity of buffer time in the range between 0.8 and 0.5, a small fluctuation in delay-coefficient arises and the increase of delay-coefficient is greater than the range over 0.8. When the value of homogeneity of buffer time is less than 0.5, timetables experience a rapid increase of delay-coefficient. For a given sight to the average value in each interval, the delay-coefficient increases over the homogeneity of buffer time.

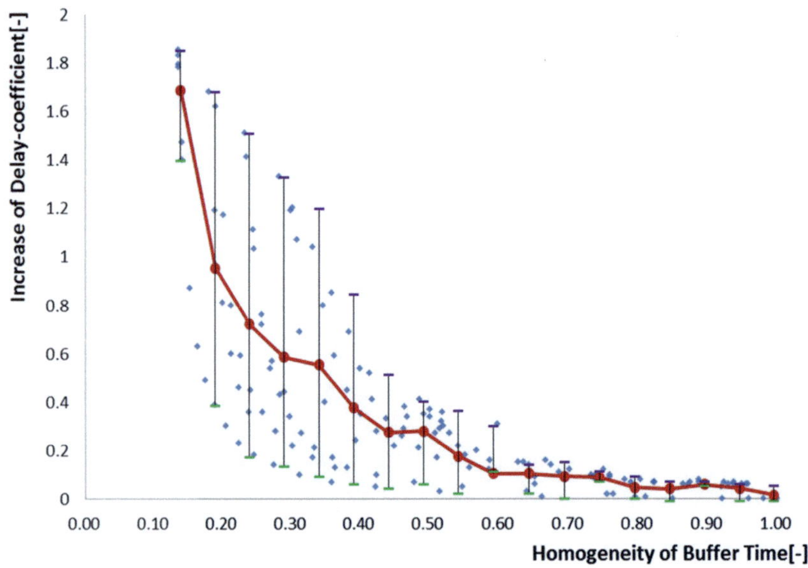

Figure 4-4:   The relationship between homogeneity of buffer time and delay-coefficient

4.3.3   Influence of Homogeneity of Running Direction

When all trains run with the same direction, the operation is homogeneous in running direction according to the new occupancy based definition. Nevertheless, there are some special occupancy elements in the railway network, like crossings and single tracks, which supports the operation of trains from different running directions. The arrangement of train movements on these elements is quite important for the capacity and quality in the whole network, since conflicts often happen on them. Constantly changing the running direction limits the capacity and hampers the operation quality. On this kind of tracks, a bundled operation, whereby separate train movements of different train types [Pachl 2013], is a homogeneous way to improve the capacity and quality. Take a single track as an example that after all trains in one direction run through the track, the other trains start to operate, and this will reach the most homogeneous situation for train runs from opposite directions regarding the running direction. The relationship between the arrangement of these trains from different directions and the operation quality is therefore investigated in this section.

The middle sections of the investigated area are such special occupation elements. The trains, from opposite directions Route 1 and Route 2 (see Figure 4-1) share these elements. In the homogeneous situation regarding running direction, 35 trains from Route 1 operate first, which are followed by 17 trains along Route 2. Therefore, there are in total 52 trains and 51 pairs of train running. At the same time, the minimum buffer times between identical trains are equally spread as 02:32. The first train departs at 0:01:37 along Route 1 and the last train departures at 5:46:47 along Route 2. At this point in time, the two trains are kept unchanged to define the investigated time period.

Based on this most homogeneous timetable, the homogeneity of running direction is changed through changing the order of train movements with different directions. Table 4-4 gives some examples of timetables with different levels of homogeneity of running direction. In order to minimize the influence of buffer time, the train movements are always distributed.

As it can be observed in Table 4-4, the graph shows that the delay-coefficient tends to increase when the homogeneity of running direction decreases. However, the difference is not so large especially for the best operation quality, which is reached for each level of homogeneity of running direction with train movements evenly spread. In this scenario, the scatter points are mainly concentrated in the region where the homogeneity of running direction is greater than 0.5 due to the layout of the investigated infrastructure.

Table 4-4:   Scenarios of heterogeneous timetables through changing sequence

Timetable	Train Sequence		Amount	Buffer Time**	Homogeneity of Operating Programs		
	Previous	Later		[s]	HBL[-]	HBU[-]	HRD[-]
Case 1	Route 1	Route 1	34	152	0.93	0.71	0.94
	Route 1	Route 2	1	4342			
	Route 2	Route 1	0	/			
	Route 2	Route 2	16	152			
Case 2	Route 1	Route 1	32	152	0.93	0.39	0.81
	Route 1	Route 2	3	887			
	Route 2	Route 1	2	887			
	Route 2	Route 2	14	152			
Case 3	Route 1	Route 1	28	152	0.93	0.53	0.71
	Route 1	Route 2	7	355			
	Route 2	Route 1	6	355			
	Route 2	Route 2	10	152			
Case 4	Route 1	Route 1	18	152	0.93	0.72	0.64
	Route 1	Route 2	17	152			
	Route 2	Route 1	16	152			
	Route 2	Route 2	0	152			

*: Buffer time on critical block section.

As shown in Table 4-4, the changes of running direction only happen on the occupancy elements in the middle area. On these occupancy elements, the homogeneity of running direction has a value below 0.5, like the examples of single tracks in Figure 3-15. However, on the other occupancy elements in the network, trains operate separately without any conflicts regarding the direction of train runs, giving the value of homogeneity of running direction as 1. Therefore, for the homogeneity in whole investigated area, most timetables have the value of homogeneity greater than 0.5.

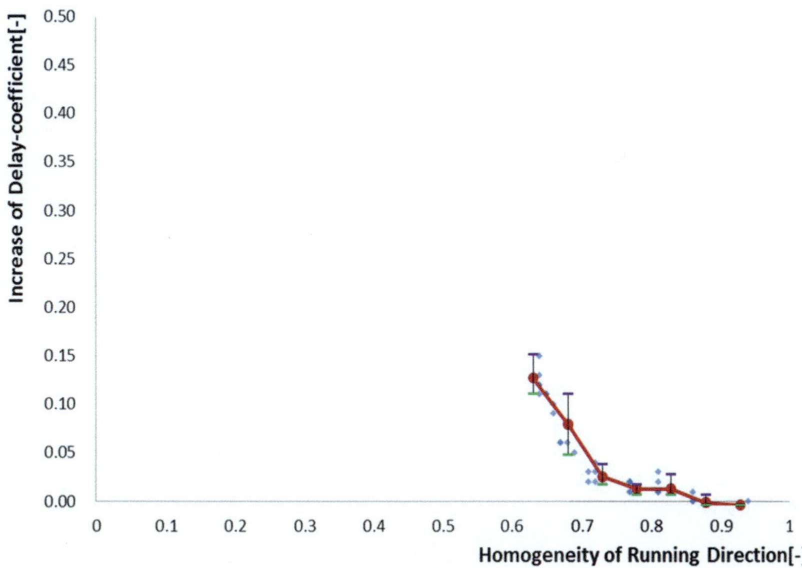

Figure 4-5:   The relationship between homogeneity of running direction and delay-
coefficient

4.4   Discussion

The homogeneity of buffer time plays an import role in the operation quality. It influences the operating quality directly and significantly. A clear negative interrelationship between homogeneity of buffer time and operating quality is presented. By contrast, the influence of homogeneity of blocking time and running direction is less strong. They influence the operating quality through the change of infrastructure occupation. When the infrastructure occupation increases, the operating quality subsequently decreases.

5 Assessment of Overall Homogeneity of Operating Programs

As introduced in Section 3, the homogeneity of operating programs can be described by homogeneity of blocking time (HBL), homogeneity of buffer time (HBU) and homogeneity of running direction (HRD). These parameters evaluate the variations in blocking time, buffer time and running direction separately. The amount of variations in blocking time, buffer time and running direction expresses the respective homogeneous magnitude that the smaller the variation is, the more homogeneous the operating program is. According to the previous analysis, each kind of variation has a negative impact on operation quality when the other factors are same, which was proved in Section 4. However, it is not comparable directly if three parameters are all different. Therefore, the objective of this section is to develop a comprehensive overall evaluation indicator of homogeneity, integrating variations in blocking time, buffer time and running into account with the consideration of their effectiveness and influence in railway operation.

5.1 Determination of Overall Homogeneity

According to the occupancy based definition of homogeneity, the variations in blocking time, buffer time and running direction make an operating program to be inhomogeneous. Therefore, homogeneity, as an indicator evaluating the structure of operating program, should consider all these variations together. Up to now, these variations are evaluated respectively through parameters the homogeneity of blocking time (HBL), the homogeneity of buffer time (HBU) and the homogeneity of running direction (HRD). Three parameters of homogeneity are integrated as overall homogeneity to have a comprehensive and comparative evaluation of operating programs.

A three-dimensional rectangular coordinate system is generated, as shown in Figure 5-1. The three normalized dimensions are HBL, HBU, as well as HRD, which can be calculated based on the new occupancy based method. The scatters present different scenario in railway operation. Scenario (a) is an absolute homogeneous operating program that same type of train evenly spread in the network. In this case, the values of HBL, HBU, and HRD are 1, indicating homogenous operation regarding blocking time, buffer time and running direction. Even slight deviations in train types and/or headways make the operation to be more inhomogeneous, like scenario (b). Conflicts between train movements with different running directions make the value of HRD to be smaller. In scenario (d), different types of trains with different origins and destinations are required to satisfy various service demands. These trains share the infrastructure leads to inhomogeneity of the three parameters. In this coordinate system, the values of HBL, HBU, and HRD determine the position of scenarios, describing the level of homogeneity.

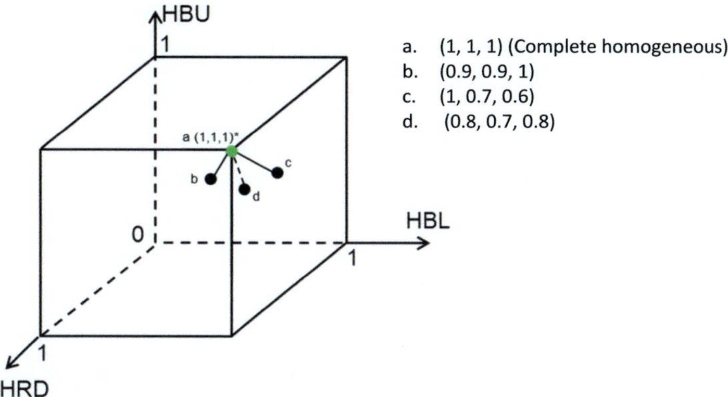

*(x, y, z): x, y, z presents the value of HBL, HBU and HRD.

Figure 5-1:   Scenarios of overall homogeneity depending on three parameters of homogeneity

Overall homogeneity of operating programs is therefore calculated as the Euclidean distance of homogeneity of blocking time, homogeneity of buffer time and homogeneity of running direction to comprehensive evaluate the variations in blocking time, buffer time and running direction. In real railway operation, a complete homogeneous operating program maybe exists in some metro networks. However, the absolute inhomogeneous operating program hardly even exists in reality. Therefore, the overall homogeneity of an operating program, expressed as Hom is calculated as the Euclidean distance to the complete homogeneous state, that each parameter of homogeneity has a value of 1. Shorter distance to the complete homogeneous state figures a more homogeneous situation.

The three parameters of homogeneity to the different extent affect the homogeneity of operating programs. Therefore, weighted Euclidean distance is utilized which gives different parameters of homogeneity different weights. It can be calculated by following formula:

$$Hom = 1 - \sqrt{w_1 * (HBL - 1)^2 + w_2 * (HBU - 1)^2 + w_3 * (HRD - 1)^2} \qquad \text{Formula (24)}$$

$$w_1 + w_2 + w_3 = 1 \qquad \text{Formula (25)}$$

where,

w_1 is the weight for homogeneity of blocking time;

w_2 is the weight for homogeneity of buffer time;

w_3 is the weight for homogeneity of running direction.

The value of overall homogeneity implies the difference between a certain operating program and its corresponding complete homogeneous operating program. High overall homogeneity means this difference is small, in other words, the operating program is relatively homogeneous.

Likewise single parameter of homogeneity, the overall homogeneity also has the domain range from 0 to 1.The value 1 of Hom represents a complete homogeneous operating program, including blocking time, buffer time and running direction, like the Scenario (a) in Figure 3-5 as an example. While an operating program has a very inhomogeneous structure if the value of homogeneity approaches 0. It should be pointed out that the overall homogeneity of an operating program could be calculated for both a particular occupancy element and a whole investigated area. Therefore, the final objective is to find suitable weights for the three parameters of homogeneity.

The first step is to identify the influence factors in regard to homogeneity of operating programs. As introduced in the variable operating program structures, the factors are classified into 4 types. According to previous research, the train length, the scheduled dwell time and mix, as well as the temporary distribution of the train path influence the homogeneity of operating programs. In the research, the developing of overall homogeneity evaluation requires a series of timetables based on well-designed experiments. Therefore, the experiments were designed related to these influence factors. Within the process of experimental design, a number of timetables with different levels of homogeneity are supposed to generate. The parameters of homogeneity (HBL, HBU, and HRD) quantify the variations in blocking time, buffer time and running direction, reflecting the homogeneity of operating programs to some extent. The HBL, HBU as well as HRD will be calculated for each timetable based on the new occupancy based method. These data can then be further used to assess the weights to three parameters of homogeneity. In this dissertation, the overall homogeneity is developed for the operation in a commuter rail network and a mixed traffic operation network (reference example).

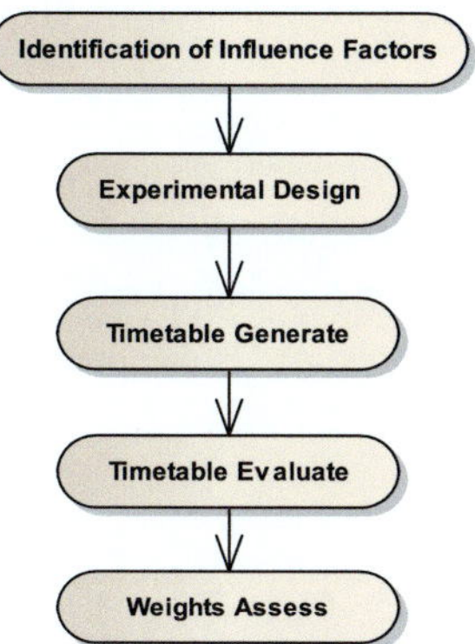

Figure 5-2:   Development process of overall homogeneity

5.2   Experimental Design and Simulation Analysis

The individual influence of each parameter of homogeneity on operation quality was investigated in Section 4 based on a commuter rail network. Abundant timetables samples with different levels of HBU and HBL were established during the research process, through the measures, e.g. alter train length, modify dwell time and shift train path in the investigated time period. The framework of this process can be found in Appendix II. The timetables are also used here to determine the weights for the parameters of homogeneity in calculate the overall homogeneity of operating programs in a commuter rail network.

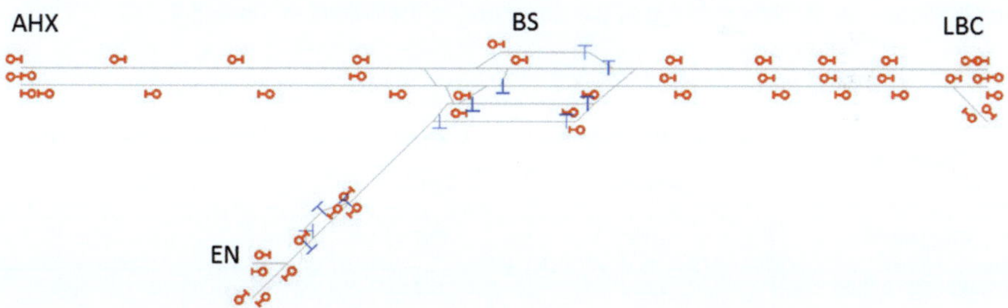

Figure 5-3:   The layout of the infrastructure network of the investigated area

In this section, the weights are also determined for a mixed traffic network (reference example). Frist of all, the framework of this reference example should be introduced in detail. The network is a theoretical network, which consists of 4 stations (named AHX, BS, LBC, and EN)

and 68 occupancy elements in total (see Figure 5-3). The tracks between station EN and station BS are single tracks, and the others are double tracks.

Three types of trains are involved in this example: long distance passenger trains (FRZ), short distance passenger trains (NRZ) and freight trains (GV). Four running directions are defined: from station AHX to station LBC, from station LBC to station AHX, from station EN to station LBC and from station LBC to station EN. The basic traffic demand is presented here as follow:

Table 5-1:   Characteristics of involved Lines

Line	Train Type	Stop Station	Dwell Time [s]
AHX-LBC	FRz	Nein	0
	NRz	Nein	0
LBC-AHX	FRz	BS	30
	NRz	BS	30
EN-LBC	NRz	EN	30
		BS	30
	GV	BS	600
LBC-EN	NRz	EN	30
		BS	30
	GV	BS	600

The characteristic of three train types are described in the following Table 5-2.

Table 5-2:   Characteristics of involved train types

Train Type	Maximum Speed [km/h]	Length [m]	Regular recovery time [%]
FRz (42/IC)	200	256	5
NRz (1074/RB)	120	122	3
GV (5558/GV)	100	669	4

A three-level, full factorial design is used in this study to obtain various timetables with different homogeneous level. Based on the variable operating program structure, train length and dwell time at stations will influence amount of blocking time on an occupancy element. Meanwhile, the train mix is one of the main factors influencing the structure of operating programs. Therefore, the experimental design for the reference example is based on these factors. Following table shows the three-levels of each factor for the overall homogeneity of operating programs in the reference example.

Table 5-3:   Levels for homogeneous factors of the reference example

Factors		Train type	Level 1	Level 2	Level 3
Length [m]		FRz	256	283	309
		NRz	122	148	174
		GV	650	994	1300
Dwell Time [s]	BS	NRz	30	60	120
	EN	NRz	30	60	120
Factors		Scenario	Scenario 1	Scenario 2	Scenario 3
Train Mix [-]		FRz:NRz:GV	1:2:1	1:3:0	0:4:0

For the first scenario of train mix (FRz: NRz: GV=1:2:1), a total of 243 (=3^5) timetables have been created. Second scenario of train mix (FRz: NRz: GV=1:3:0) contains totally 81 (=3^4) timetables have been created due to the absent of freight trains (GV). Similarly, the train length of long distance passenger trains (FRz) is not considered in the third scenario of train mix. Therefore, third scenario of train mix demands 27(=3^3) timetable. As a result, a total of 351 timetables (=243+81+27) have been created to develop the weighted Euclidean distance model for overall homogeneity evaluation.

5.3   Weights Assessing

Due to the specific characteristics of railway operation, the values of parameters of homogeneity have their own special range and distribution type in different railway networks, which raise different weights for three parameters of homogeneity.

Considering the distribution of values of different parameters, if the parameter has less fluctuation, it will be given higher weight. For example, trains are normally separated from other trains with opposite running direction in the commuter rail network. Therefore, no conflicts will arise due to the different running directions, which make value of HRD to be 1. In this case, if there is one train runs from opposite direction, due to the restrictions on infrastructure or some dispatching reasons, the influence on the homogeneity will be huge.

In this dissertation, the concept of entropy was used to determine the weights of parameters in the weighted Euclidean distance to evaluate the overall homogeneity of operating programs. Entropy is a measure of disorder or randomness of a system. Then entropy is high, when the difference of the value among the evaluating objects on the same indicator is huge. On the other hand, if the evaluating objects have determined value on the same indicator, the entropy is smallest as 0.

The steps for weights determination of parameters of homogeneity using entropy method are as follows. The entropy of jth parameter of homogeneity is defined as:

$$S_j = -\left(\sum_{i=1}^{m} p_i^{(j)} \ln p_i^{(j)}\right) / \ln m \qquad \text{Formula (26)}$$

Where,

$p_i^{(j)}$ is the probability of the value of jth parameter of homogeneity in i^{th} section.

Since the parameters of homogeneity are continuous variables, the values are divided into 20 sections with interval as 0.5.

$$p_i^{(j)} = y_i^{(j)} / \sum_{i=1}^{m} y_i^{(j)} \qquad \text{Formula (27)}$$

Where,

$y_i^{(j)}$ is the amount of values in j^{th} section the parameter of homogeneity j, $i \in [1, 20]$

m is the total number of timetables.

The indicator S_j influences the weight coefficient of parameters of homogeneity. In other words, the less disordered the timetables for the parameters of homogeneity j; the change of homogeneous j has more importance for the overall homogeneity evaluation. In Therefore, the weight for the parameter of homogeneity j is thus given by

$$w_j = (1 - S_j) / \sum_{j=1}^{n} (1 - S_j) \qquad \text{Formula (28)}$$

Where,

S_j is the entropy of jth parameter of homogeneity,

n is the total number of parameter of homogeneity.

5.3.1 Weights for Commuter Rail Networks

The weights obtained by entropy method are 0.31 for HBL, 0.24 for HBU and 0.45 for HRD (see Table 5-4). According to the weight coefficient, the parameter of homogeneity HRD is most important to evaluate the homogeneity in commuter rail network. For example, the situation that a train from opposite direction added into a railway system, which directly changes the value of HRD, has a big influence on the overall homogeneity. The weight of HBL is relative higher than the weight of HBU, which may be the result of same speed for all trains operated in the network.

Table 5-4: Divergent and Weights of Homogenous Parameters for the Commuter Rail Network

Parameters of homogeneity	Entropy	Weights
HBL	0.3108	0.31
HBU	0.4683	0.24
HRD	0	0.45

Therefore, the overall homogeneity of operating programs in the commuter rail network can be calculated with following weighted Euclidean Distance:

$$Hom = 1 - \sqrt{0.31 * (HBL - 1)^2 + 0.24 * (HBU - 1)^2 + 0.45 * (HRD - 1)^2} \qquad \text{Formula (29)}$$

5.3.2 Weights for Reference Example

For the reference example, the weights for HBL, HBU and HRD are respectively 0.31, 0.34 and 0.35 by the entropy method. In this kind of network, three parameters of homogeneity are calculated almost of equal importance. The weight of HRD in the mixed traffic network is obviously smaller than that in the commuter rail network. Meanwhile, the parameter of homogeneity HBU is more important in the mixed traffic network.

Table 5-5: Divergent and Weights of Parameters of Homogeneity for the Reference Example

Parameters of homogeneity	Entropy	Weights
HBL	0.1956	0.31
HBU	0.1332	0.34
HRD	0.0873	0.35

The overall homogeneity of operating programs can be evaluated with following formula for the reference example.

$$Hom = 1 - \sqrt{0.31 * (HBL - 1)^2 + 0.34 * (HBU - 1)^2 + 0.35 * (HRD - 1)^2} \qquad \text{Formula (30)}$$

5.4 Influence of Overall Homogeneity

5.4.1 Commuter Rail Network

In Section 4, an amount of timetables are generated to investigate the influence of each parameter of homogeneity on operation quality. The overall homogeneity is computed for these timetables here to further find out its relationship with operation quality. The relationship between overall homogeneity and delay-coefficient in the commuter rail network is depicted in the following scatter plot (see Figure 5-4). The scatters are distributed in the whole range from 0.5 to 1. This figure indicates a remarkable trend that the delay-coefficient increases as the overall homogeneity decreases.

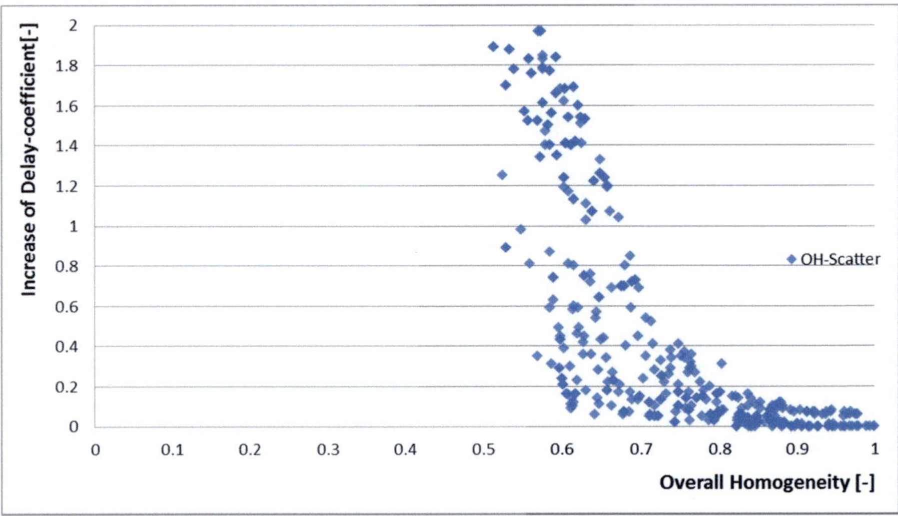

Figure 5-4: The relationship between overall homogeneity and delay-coefficient in the commuter rail network

This influence can be divided into several phases. At first, the average delay-coefficient increases at a very slow rate, as the overall homogeneity deteriorate. In addition, the deviation of delay-coefficient for each value of homogeneity is relatively small. And then the operation quality steps in a fluctuating stage, in which the deviation in operation quality of each homogeneity value is greater. In the third stage, the average value of delay-coefficient in each interval increased rapidly with the decrease of overall homogeneity. But, the deviation of each scatter is quite great. Some of the timetables in this stage possess bad operation quality. Nevertheless, some timetables still have quite good operation quality. It reveals that, for the timetables in this interval, they also can have a good operation quality through good arrangement of trains.

5.4.2   Reference Example

For the reference example, the influence of the overall homogeneity on the delay-coefficient is shown in Figure 5-5. The values of the overall homogeneity of these timetables are always over 0.7, which is because of the limitations of train length and restrictive investigated time period.

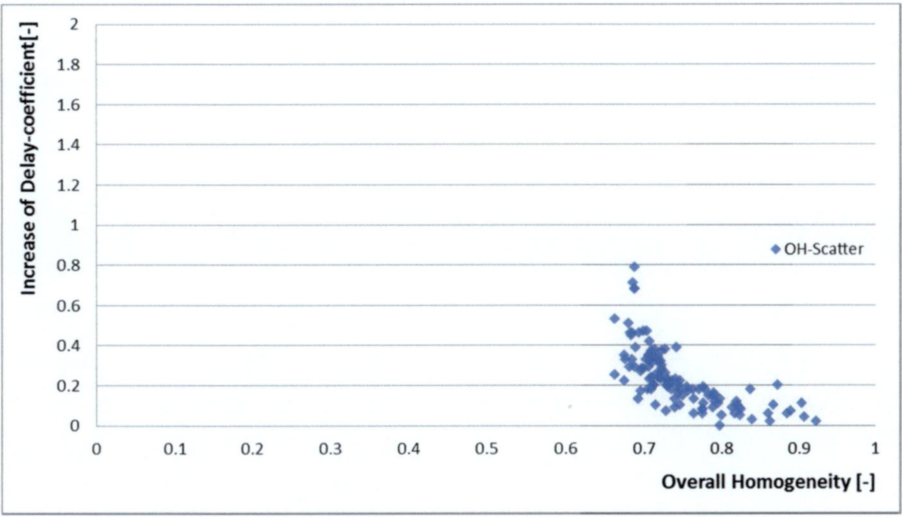

Figure 5-5:   The relationship between overall homogeneity and delay-coefficient in the mixed traffic network

As it can be observed, the scatter plots visually demonstrate the increase of delay-coefficient over the overall homogeneity in the reference example. In general, the delay-coefficient tends to increase along with the decrease in overall homogeneity. But the change of delay-coefficient is relatively small especially for timetables with a high value of overall homogeneity over 0.8. While, along with the continuous decrease of overall homogeneity, the increase of delay-coefficient is huger.

5.5   Discussion

The variations in blocking time, buffer time and running time can be evaluated by parameters of homogeneity HBL, HBU and HRD, which further be reflected in the value of overall homogeneity. In different networks, three parameters of homogeneity were given different weights in overall homogeneity evaluation based on the weighted Euclidean distance model. The study shows that HRD is the most important parameter for the commuter rail network. In the mixed used traffic network, the three homogenous parameters are almost of equal importance. The importance of HRD is less in the mixed used traffic network comparing to the commuter rail network.

The overall homogeneity gives a synthetic and integrated evaluation of the structure of operating programs, which is able to distinguish different levels of homogeneity. It enables a comparison between any operating programs, even if the values of each parameter of homogeneity are different. At the same time, operation quality is proved to be worse when the overall homogeneity is smaller.

6   Influence of Homogeneity in Capacity Research

As mentioned before, the efficiency of railway operation includes two aspects. Firstly, the infrastructure should be fully utilization with a certain amount of trains. Secondly, the operation quality fulfills the expected level of operators and customers. Therefore, the efficient utilization of infrastructure is a balance of capacity and operation quality through perfect arrangement of diversified services.

The negative relationship between operation quality and capacity of an infrastructure with a given operating program could be depicted visually by a waiting time function. As shown in Figure 6-1 the magnitude of unscheduled waiting time depends on the number of trains in the network and the homogeneity of operating program.

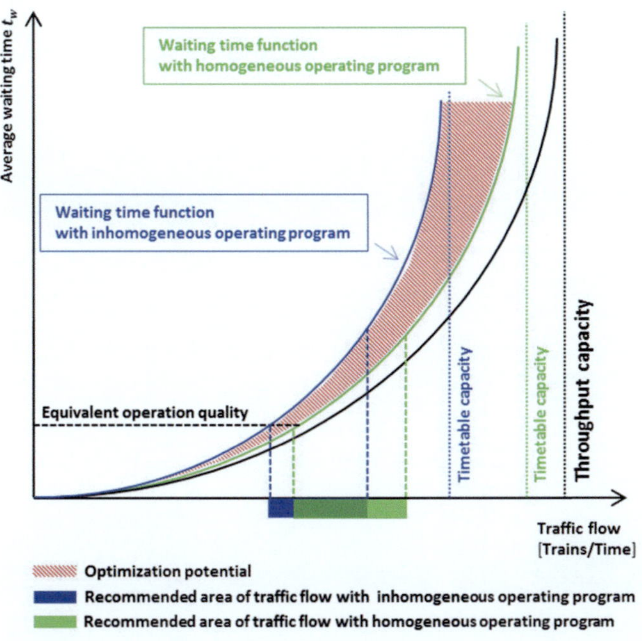

Figure 6-1:   Influence of homogeneity of operating programs on waiting time function

Pachl mentioned that the distance between the timetable capacity and the throughput capacity based on the maximum theoretical capacity is affected by the homogeneity of operating programs [Pachl 2013; Chu 2014]. In the case of very homogeneous timetable, which are mainly parallel operated trains, a much higher scheduled utilization of the capacity of the infrastructure is possible than that for a timetable with an inhomogeneous structure. On the other hand, inhomogeneous timetables offer the possibility to cope with the disturbances at the expense of eliminating the capacity.

The objective of this chapter is to check the performances of operating programs with different levels of homogeneity over traffic flow, which can be described by the curve of the waiting time function. As the traffic flow increases, the change of values of homogeneity is one study point. Additionally, with different traffic flows, the influences not only of single parameters of homogeneity but of overall homogeneity on the relationship between unscheduled waiting time and traffic flow are investigated.

6.1   Case Study

Figure 6-2 describes the work flow to analyze the influence of homogeneity of the operating program on the average waiting times with different traffic flow and the shape of waiting time function with different level of homogeneity.

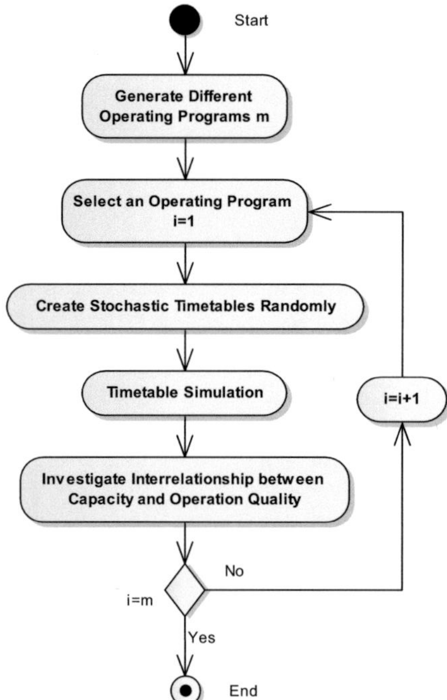

Figure 6-2:   Algorithm to analyze the influence of homogeneity in capacity research

In this analysis, random timetables are generated based on a set of operating programs. These operating programs have different train mix; therefore the variations in these trains are different for these operating programs.

Each operating program scenario is used to generate random timetables by stepwise increasing of the train density while retaining the structure of the base schedule. Timetable simulation will then be conducted to each random generated timetable to simulate the train runs, meanwhile, recording the time points of each train through each block section. Based on this protocol, each parameter of homogeneity and the overall homogeneity can be calculated for these stochastic timetables based on the occupancy based method. Besides, unscheduled waiting time is recorded as the indicator of operation quality.

In total, 5 scenarios of operating program are generated with different train mix for the commuter rail network, which was detailed described before in section 4.1. In these scenarios, only trains with route 1 were considered in operation. These operating programs were compressed specifically based on these operating programs from 10% to 200% with an interval as 5%. For each timetable, the homogeneity of blocking time and homogeneity of buffer time were calculated, and the homogeneity of running direction kept as 1 for all timetables. The overall homogeneity was computed through the corresponding formula for commuter rail network as described in Chapter 5.

6.2   Analysis of Results

Figure 6-3 presents the interrelationship between unscheduled waiting time and traffic flow of different operating program scenarios. Since these operating program scenarios are a roughly description of the train mix and the characteristic of each train type. They don't contain the temporal information of train movements, e.g. headways and buffer times. Therefore, the homogeneity of blocking time is used to describe the degree of homogeneity of operating programs. The random generated timetables based on the same operating program scenario have almost same value of homogeneity of blocking time.

The results confirm that the operation quality deteriorate with increased traffic flow. The rule is proved by these 5 operating program scenarios. But the performance of this relationship is different for each operating program scenario. An operating program with the homogeneity of operating program can have better operation quality when the traffic flow is same. With the same requirement of unscheduled waiting time, the more homogeneous operating program allows more trains operated in the network.

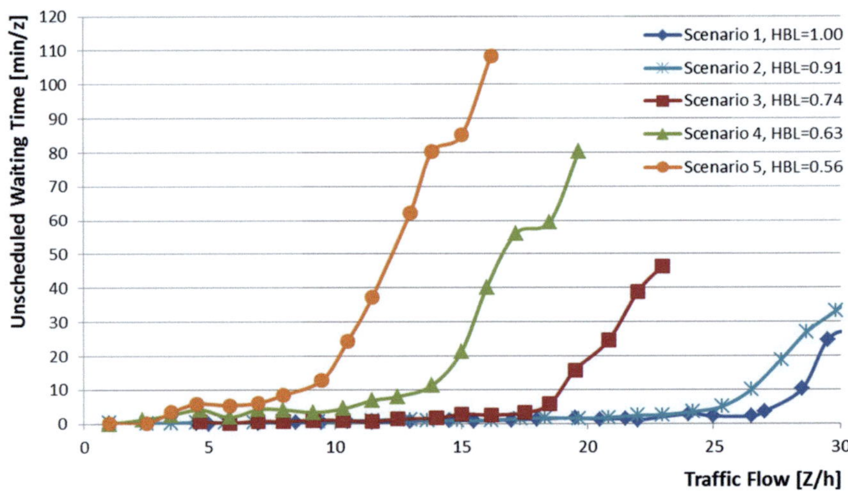

Figure 6-3:   Capacity research of different operating program scenarios

The influence of homogeneity of blocking time (HBL) with different traffic flow is shown in following Figure 6-4. The values of HBL range from an inhomogeneous value of 0.45 to the absolute homogenous value of 1. Due to the continuity of HBU, the timetables are classified into 20 groups based on the value of HBL with homogeneous interval as 0.05. In each interval, timetables have almost same level of homogeneity of blocking time.

It is founded that the average unscheduled waiting times decrease with the increase of HBL for timetables having same traffic flow. In other words, when the traffic flow is pre-determined, a better operation quality (less average unscheduled waiting time) can be achieved with greater value of HBL. However, for timetables operate fewer trains, the influence of HBU is relatively small. The difference of average waiting time between high HBL and low HBU is more remarkable, when the traffic flow is growing up. If the average waiting time is kept constant, the capacity expands through homogeneous blocking time. Therefore, more capacity can be provided through homogenizing blocking time.

Beyond that, for each interval of HBL, the average waiting time has a growing trend with the traffic flow as shown in Figure 6-4. The average waiting time increases rapidly after a certain point. Along with the decrease of HBL, the critical point shift to left. And the smaller is the HBL, the steeper is the waiting time function, which valid the research of Pachl [Pachl 2013].

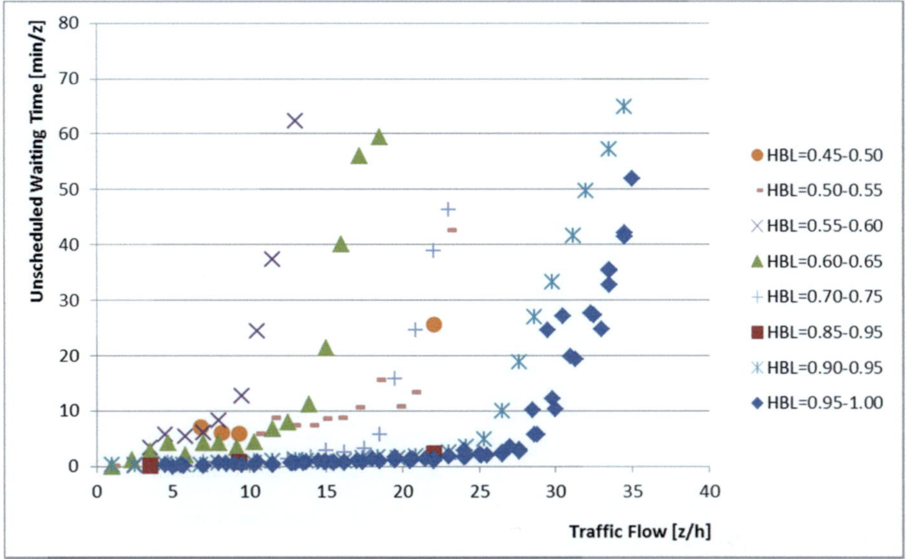

Figure 6-4:   Influence of homogeneity of blocking time in capacity research

The same timetables are also used to check the influence of homogeneity of buffer time (HBU) with different traffic flows. The lowest value of HBU is almost 0, and the highest value is 1, indicating an evenly distribution of buffer times. Similar to the analysis regarding homogeneity of blocking time, the timetables are likewise classified into 20 groups based on the value of HBU with homogeneous interval as 0.05. The analysis of result is presented in Figure 6-5.

It is founded also that the average waiting times increase with the traffic flow. For the timetables in the same interval, the single points create a negative exponential tendency. In general, the curve with lower homogeneity of buffer time is above the curve with higher homogeneity of buffer time. It means that the operation quality of timetables having lower HBU is generally worse with more average waiting time. When the traffic flow is the same, higher value of HBU means smaller waiting time, which also valid the result in Section 4.3.

The cross point of the line with certain operation quality and the curve is full of meaning. The point means that, for this traffic flow, the requirement of operation quality can be achieved with this value of HBU. When the traffic flow is smaller, the corresponding value of HBU of this cross point is bigger. In other words, along with the decrease of traffic flow, a homogeneous buffer time is required to have a certain operation quality. Therefore, in the area of low traffic flow, the influence of homogeneity of buffer time is significant.

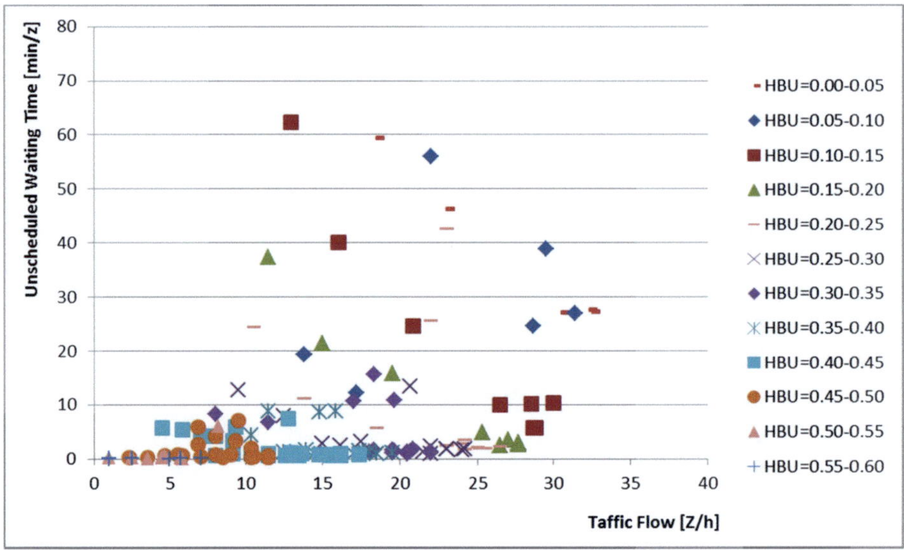

Figure 6-5:   Influence of homogeneity of buffer time in capacity research

As the number of trains in operation increases, it is also more difficult to cope with delays. This means that even small delays can have a major impact on the entire network. Furthermore, offering new or additional services on the existing infrastructure with existing operation methods may be limited or not even possible because available capacity is already fully utilized.

The overall homogeneity (OH) is calculated as the Euclidean distance of HBL, HBU and HRD to the absolute homogeneous situation. The most inhomogeneous timetables have the value of OH less than 0.35. The most homogenous timetables have the value of OH more than 0.7. Similarly, these timetables are divided into 20 groups based on the value of OH with homogeneous interval as 0.05. The result is presented in Figure 6-6.The result also reveals the law that operation quality intends to be worse when the traffic flow increases.

In general, the curve with lower homogeneity of buffer time is above the curve with higher homogeneity of buffer time. It means that the operation quality of timetables having lower HBU is generally worse with more average waiting time. When traffic flow is same, higher value of HBU means smaller waiting time, which valid the results in Section 4.3, too.

For the timetables in the same homogeneous interval of OH, the single points also create a negative exponential tendency, which is similar to the influence of homogeneity of buffer time. For the same homogeneous interval of OH, the higher traffic flow has better operation quality;

on the contrary, the operation quality is worse for the timetables having less traffic flow. It means that the requirement of overall homogeneity is higher when the traffic flow is low.

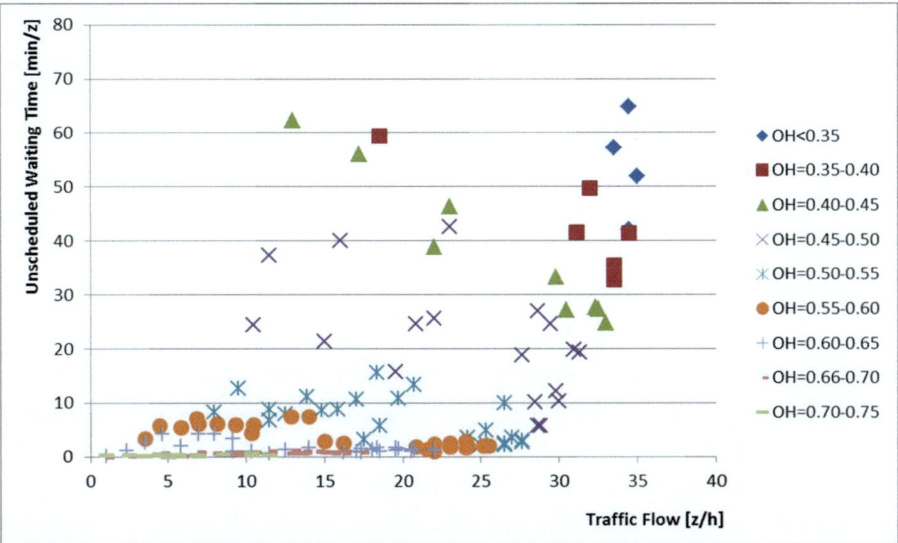

Figure 6-6:   Influence of overall homogeneity in capacity research

As the number of trains in operation increases, it is also more difficult to cope with delays even with good homogeneity of overall homogeneity. This situation is same for the homogeneity of buffer time.

7   Summary, Conclusion and Future Development

In this thesis, an expansion of existing definitions of homogeneous operation is developed from the new perspective of infrastructure occupancy, which is significant for efficient utilization. According to the new occupancy based definition, railway operation is considered to be complete homogeneous when there are no variations in blocking time, buffer time and running direction. Blocking time together with buffer time depicts the occupancy of infrastructure. The running direction is a supplementary indicator to describe the occupancy of infrastructure, in particular for stations, points, crossings, single-track lines and double-track lines with bidirectional operation.

Based on the new occupancy based definition, the homogeneity of operating programs is evaluated by three parameters: homogeneity of blocking time (HBL), the homogeneity of buffer time (HBU) and the homogeneity of running direction (HRD). In the calculation process, the coefficient of variation is used to quantify the variations in blocking time, buffer time and running direction. This method is the first attempt to evaluate the homogeneity of an operation program considering the occupancy of infrastructure. It is proved that the three parameters can well quantify homogeneous level. The evaluation can, furthermore, be conducted not only on a particular occupancy element but in an entire investigated area, which is reported in the literature for the first time. In addition, the developed new occupancy based method is independent of infrastructure types and characteristic of trains. For the calculation of the parameters, only the infrastructure information and timetables are required with the assistance of simulation software and evaluation software. It is comprehensive as well as universally applicable in any infrastructure model.

The indicator overall homogeneity, which is able to evaluate the homogeneity of operating programs comprehensively, is generated combining these three parameters of homogeneity. The overall homogeneity of operating program is computed as the Euclidean distance of the weighted homogeneity of blocking time, buffer time and running direction to the absolute homogeneous state. In this research, the homogeneity of blocking time, buffer time and running direction are assigned different weights for the commuter rail network and the mixed traffic network, based on the entropy of single parameter of homogeneity. Up to now, the weighted overall homogeneity can be used to distinguish different levels of homogeneity of operating programs developed in this thesis.

The influence of each homogenous parameter on operation quality was investigated for a real commuter rail network. The homogeneity of buffer time plays an import role in the operation quality. It influences the operating quality directly and significantly. A clear negative interrelationship between homogeneity of buffer time and operating quality is presented. By contrast, the influence of homogeneity of blocking time and running direction is less strong. These two parameters of homogeneity affect the operating quality through the change of infrastructure occupation. When the infrastructure occupation increases, the operating quality subsequently decreases. A negative relationship between overall homogeneity and operation quality is also proved in this thesis.

In addition, the influence of homogeneity of operating programs differs for traffic flows. For lower traffic flow, the homogeneity of blocking time is of less importance, that even timetable with considerable variation in blocking time of train runs can also have a good operation quality. But, the difference between timetables with high value and low value of homogeneity of blocking time is huge. In this case, the arrangement of buffer time between these trains plays a more important role. But for high traffic flow, the operation quality is more sensitive to the homogeneity of blocking time.

To summary, this system is significant for real rail operation considering the operational use of existing infrastructure. The parameters of homogeneity generated a system to evaluate the timetables regarding the occupancy of infrastructure, which has not been studied so far. They also can be used as indicators in identifying timetables with a high efficient utilization of existing infrastructure. Besides, it helps timetable planners to track down the weakness in timetables based on the value of homogeneity. The corresponding measures to deal with different type of weaknesses can also be investigated in future research.

In future, the three parameters can be assigned different weights to adapt various scenarios, based on either the individual influences on the overall homogeneity or special requirements from infrastructure managers. The weighted Euclidean distance makes the overall homogeneity of operating programs more flexible and adaptable for the future study.

As mentioned above, the homogeneity of operating programs in railway systems is distinguished in planning and operation process. Due to various disturbances, the railway operation deviates from the scheduled timetable. Both the homogeneity in planning and operation process can be calculated with the new occupancy based method. It is an incentive to investigate which factors influence the difference between homogeneity level in operating and in planning in future.

8 Anhang I: Introduction of Enterprise Architect

The unified modeling language (ULM) is a graphical purposed-oriented modeling language for visualizing, specifying, constructing and documenting the artifacts of a software-intensive system. The UML offers a standard way to construct a system´s blueprints, including conceptual things such as business processes and system functions as well as concrete things such as programming language statements database schemas, and reusable software components.

Enterprise architect is visual modeling and design tool based on the ULM. It supports the design and construction of software systems, modeling business processes, and modeling industry based domains. In this dissertation, enterprise architect modeled the algorithm and workflow used in the research.

Notations are the most important elements in modeling. Efficient and appropriate use of notation is very important for making a complete and meaningful model. Different notations are available for things and relationships.

- **Things:**

Graphical notations in structural things are most widely used. Following are detailed descriptions of the elements commonly used when modeling with UML Structure Diagrams in Enterprise Architect.

Name	Notation	Definition
Class	Class	A class is a representation of a type of object that reflects the structure and behavior of such objects within the system. It can have attributes (data) and methods (operations or behavior).
Interface	«interface» Interface	An interface is a specification of behavior (or contract) that implementers agree to meet.
Component	Component	A component is a modular part of a system, whose behavior is defined by its provided and required interface.

Collaboration		Collaboration defines a set of cooperating roles and their connectors, which are used to collectively illustrate a specific functionality.
Node		A node is a physical piece of requirement on which the system is deployed, such as a workgroup server or workstation.
Object		An Object is a particular instance of a Class at run time. It is presented in the same way as the class. The only difference is the name which is underlined as shown in the following figure.
Package		A Package is a namespace as well as an element that can be contained in other Package's namespaces. A Package can own or merge with other Packages, and its elements can be imported into a Package's namespace.
Actor		An Actor is a user of the system which can be a human user, a machine, or even another system or subsystem in the model. Anything that interacts with the system from the outside or system boundary is defined as an Actor. Actors are typically associated with Use Cases.
Note		A note element is a textual annotation that can be attached to a set of elements of any other type. It is created separately, using a note link connector.

This section provides detailed description of the elements commonly used in modeling with UML Behavioral Diagrams[26] in enterprise architect.

---

[26] UML Behavioral Diagrams depict the elements of a system that are dependent on time and that convey the dynamic concepts of the system and how they relate to each other.

Name	Notation	Definition
Action	Action	An action element describes a basic process or transformation that occurs within a system and is the basic functional unit within an activity diagram.
Actor	Actor	An Actor is a user of the system which can be a human user, a machine, or even another system or subsystem in the model. Anything that interacts with the system from the outside or system boundary is defined as an Actor. Actors are typically associated with Use Cases.
Collaboration	Collaboration	Collaboration defines a set of cooperating roles and their connectors, which are used to collectively illustrate a specific functionality.
Decision/Choice		A decision is an element of an activity diagram or interaction overview diagrams that indicates a point of conditional progression. The choice pseudo state is used to compose complex transitional paths in, where the outgoing transition path is decided by dynamic, run-time conditions. The run-time conditions are determined by the actions performed by the state machine on the path leading to the choice.
Initial		In Activity diagrams, it defines the start of a flow when an Activity is invoked. With State Machines, the Initial element is a pseudo state used to denote the default state of a Composite State.
Final		There are two nodes used to define a Final state in an Activity, both defined in UML 2.5 as of type Final Node.
Use Case	Use Case	A Use Case is a UML modeling element that describes how a user of the proposed system interacts with the system to perform a discrete unit of work. It describes and signifies a single interaction over time that has meaning for the end user (person, machine or other system), and is required to leave the system in a complete state: the interaction either completed or rolled back to the initial state.

- **Relationship**

A model is not complete unless the relationships between elements are described properly. The relationship gives a proper meaning to a model. Following are the different types of relationships defined in EA.

Name	Notation	Definition
Dependency	- - - - ->	A dependency is a relationship that signifies that a single or a set of model elements requires other model elements for their specification or implementation.
Association	———	An associate implies that two model elements have a relationship, usually implemented as an instance variable in one or both classes.
Generalization	———▷	A generalization is used to indicate inheritance. It is a taxonomic relationship between a more general classifier and a more specific classifier. Thus, the specific classifier inherits the features of the more general classifier.
Realization	- - - -▷	A realization signifies that the client set of elements are an implementation of the supplier set, which serves as the specification.
Aggregation	——◇	An aggregation connector is a type of association that shows that an element contains or is composed of other elements.
Composition	——◆	A composition is used to depict an element that is made up of smaller components, typically in a Class or Package diagram.

9 Anhang II: Framework of Commuter Rail Network (S-Bahn) Case

The analysis of influence of three parameters of homogeneity was analyzed based on a commuter rail network (S-Bahn) for the reason that the operation in commuter rail network is relative simple. It is possible to operate an absolute homogeneous timetable according the definition of the occupancy-based homogeneity. This appendix provides some information of infrastructure and timetables/operating programs in details.

Firstly, the layout of the network is presented in the following Figure 9-1. This network simulates an operational S-Bahn line in practice. It is a double-track railway which consists of 8 stations in total. In practical operations, trains on different tracks will not interact with each other. Nevertheless, in order to analyze the influence coming from homogeneity of running direction, some adjustments have been made to Route 2. As shown in the figure, Route 1 and Route 2 intertwine in the middle of the network. Route 1 is composite of 14 occupation elements and Route 2 has 11 occupation elements in total.

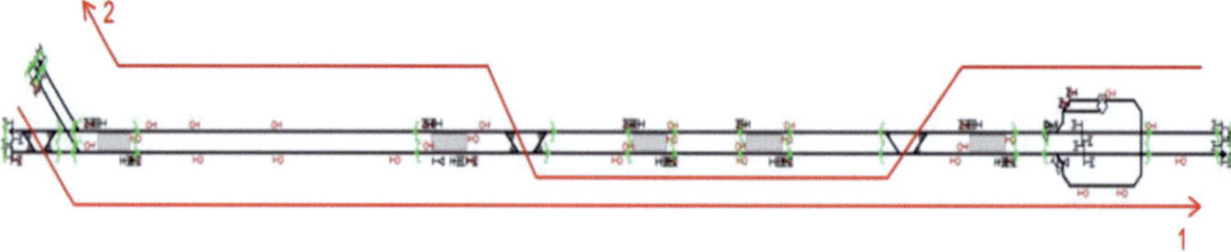

Figure 9-1:   The infrastructure network of the investigated area

As mentioned in the text, the basic princple of the influence analysis is to evalute the operation quality of timetbles with different levels of homogeneity and then figure out the interrelationship between operation quality and corresponding parameter of homogeneity. Therefore, the analysis requires a series of timetable which contains different levels of ho-mogeneity. Different measures have been conducted to the basic timetables to generate different homogeneous levels based on absolute homogeneous timetables.

- **Homogeneity of Blocking Time**

For the parameters HBL and HBU, only Route 1 was studied since the situation of the opposite direction is same. In the original homogeneous timetable, there are 69 identical trains (train 2102) run by Route 2 with headway 5 min, complying with real operating during high peak hours. The first train departs at 0:01:37 and last train at 5:41:37. To ensure that the total operational time is the same for all timetables, the first train and the last train won't make any changes, including departure time, length or dwell time.

Two measures were applied to generate various value of HBL. One measure is to alter the train length. Eight different trains are generated for the analysis. These trains have same speed and identical dynamic behaviors. However, the only difference is the train length, which ranges from 1m to 1360m. Longer trains need more time to run through the clearing points of each block section. The different train lengths make the blocking time of a train on each occupancy element varies from one another. The detailed information is presented in the Table 9-1.

Table 9-1:   Blocking times of different trains on respective occupancy elements

Train ID	1000	2000	2101	2102	2103	2105	2110	2120
Train Length [m]	1	10	67	136	207	343	688	1360
OE 1	29	29	32	36	42	104	128	161
OE 2	116	115	119	122	126	132	149	184
OE 3	57	57	60	64	67	74	90	192
OE 4	73	72	76	79	82	91	169	218
OE 5	62	62	68	122	129	140	167	270
OE 6	104	103	108	113	119	130	154	260
OE 7	74	73	78	84	139	152	181	282
OE 8	133	132	138	144	150	161	235	298
OE 9	55	55	60	67	116	129	157	263
OE 10	119	118	123	129	135	146	226	275
OE 11	61	60	65	73	128	139	166	182
OE 12	138	137	143	148	154	163	178	178
OE 13	60	60	64	66	66	66	70	70
OE 14	57	57	57	57	57	57	61	61
Sum	1138	1130	1191	1304	1510	1684	2131	2894

The first column label spanning the OE rows reads **Blocking Time [s]**.

However, the magnitude of change is limited by this method because the difference in blocking time between the longest and shortest train is not dramatic. It is not reasonable to infinitely either extend or shorten train lengths in practical process.

The blocking time in a station area always considered to be the scheduled dwell time. Therefore, the second measure is to modify the schedule dwell time of these trains on respective stations. In practice, a train may wait at a station or before a signal to avoid conflicts or ensure connection, especially for freight trains. There are nine stations in this commuter rail network. In the original homogeneous timetable, the planning dwell times are presented in Table 9-2. In order to generated change the occupancy time on each timetables. Different scenarios of scheduled dwell time are presented in Table as well.

Table 9-2:   Different scenarios of dwell time arrangement

Scenarios		1	2	3	4	5	6
	Station 1	0	0	0	0	0	0
	Station 2	36	72	180	360	480	600
	Station 3	36	72	180	360	480	600
Scheduled Dwell time [s]	Station 4	36	72	180	360	480	600
	Station 5	30	60	150	300	840	840
	Station 6	36	72	180	360	480	600
	Station 7	0	0	0	0	0	0
	Station 8	0	0	0	0	0	0

The change of homogeneity of blocking time through this method is also constricted due to the length of investigated time period. In some timetables, both actions are therefore combined to generate larger variation in blocking time.

- **Homogeneity of Buffer Time**

The influence analysis of homogeneity of buffer time on operation quality is based on the same complete homogeneous timetable on Route 1 as in the analysis of homogeneity of blocking time. In the investigated time period (from 0:00:00 to 06:00:00) 69 identical trains evenly spread with the headway as 5 min and the minimum buffer time as 152s between two paths. Similarly, the first train and the last train won't make any changes to keep the usable operation time constant.

Various inhomogeneous timetables regarding buffer time were generated through shifting train paths inside the investigated period. Shifting a train path will alter the buffer times of this train between the previous and following trains, further altering the temporal distribution of buffer time. The buffer time can reach minimum as 0 between consecutive trains, guaranteeing no conflicts.

Shifting a train path, in other words, is to adjust the departure time. Therefore, with the assistance of software PULEIV, timetables are randomly generated. Some minor adjustments should be made to avoid conflicts between train paths. These timetables mainly belong to an intermediate state, not very homogeneous and not very inhomogeneous with the value of HBL between 0.5 and 0.7. Therefore, some special situations, which are more homogeneous or inhomogeneous, were made manually complementally.

Table 9-3:   Extreme examples of buffer time distribution

Grouping	Train Group*		Buffer Time [s]	
			Relative Homogeneous	Relative Inhomogeneous
Scenario 1	2	Inside group	120	0
		Between groups	184	304
Scenario 2	5	Inside group	120	0
		Between groups	287	795
Scenario 3	10	Inside group	120	0
		Between groups	482	1723
Scenario 4	15	Inside group	120	0
		Between groups	664	2584
Scenario 5	20	Inside group	120	0
		Between groups	845	3445
Scenario 6	30	Inside group	120	0
		Between groups	1208	5168
Scenario 7	68	Inside group	120	0
		Between groups	2296	10336

First of all, the 69 trains are divided into different groups. In the relative homogeneous situation, the trains in one group are evenly spread with 120s buffer time between consecutive train paths. In the relative inhomogeneous situation, the trains in one group are compressed maximum without any buffer time. The rest buffer time are evenly distributed between groups. Scenario 7 is the most inhomogeneous situation, in which ahead 68 trains are compressed without buffer time between consecutive trains leaving a huge time gap between the last two trains. The value of HBL approaches 0.14 in this scenario.

- **Homogeneity of Running Direction**

The homogeneity of running direction is significant only when the conflict exists due to different running direction. Therefore, different from the original homogeneous timetable for HBL and HBU, another original homogeneous timetable is used to generate various HRD. In this absolute homogeneous timetable, both Route 1 and Route 2 are investigated. As known in the layout, the trains on Route 1 have potential conflict with trains on Route 2 because of the opposite running direction. In the homogeneous situation regarding running direction, 35 trains from Route 1 operate first followed by 17 trains along Route 2. Therefore, there are in total 52 trains and 51 pairs of train running. At the same time, the minimum buffer times between identical trains are equally spread as 02:32. The first train departs at 0:01:37 along Route 1 and the last train departures at 5:46:47 along Route 2. At this point in time, the two trains are kept unchanged to define the investigated time period. Therefore, there are totally 17 scenarios were generated. It was found that the number of train sequence of Route 1 to Route 2 is always 1 more than the train sequence of Route 2 to Route 1. These two amounts are interrelated. The shift of one train always related to one change of Route 1 to Route 2 and one change of Route 2 to Route 1.

Table 9-4:   Different scenarios of train sequence arrangement

Scenarios	Train Sequence		Amount [-]
	Previous	Later	
Scenario 1	Route 1	Route 2	1
	Route 2	Route 1	0
Scenario 2	Route 1	Route 2	2
	Route 2	Route 1	1
Scenario 3	Route 1	Route 2	3
	Route 2	Route 1	2
Scenario 4	Route 1	Route 2	4
	Route 2	Route 1	3
Scenario 5	Route 1	Route 2	5
	Route 2	Route 1	4
Scenario 6	Route 1	Route 2	6
	Route 2	Route 1	5
Scenario 7	Route 1	Route 2	7
	Route 2	Route 1	6
Scenario 8	Route 1	Route 2	8
	Route 2	Route 1	7

Scenario 9	Route 1	Route 2	9
	Route 2	Route 1	8
Scenario 10	Route 1	Route 2	10
	Route 2	Route 1	9
Scenario 11	Route 1	Route 2	11
	Route 2	Route 1	10
Scenario 12	Route 1	Route 2	12
	Route 2	Route 1	11
Scenario 13	Route 1	Route 2	13
	Route 2	Route 1	12
Scenario 14	Route 1	Route 2	14
	Route 2	Route 1	13
Scenario 15	Route 1	Route 2	15
	Route 2	Route 1	14
Scenario 16	Route 1	Route 2	16
	Route 2	Route 1	15
Scenario 17	Route 1	Route 2	17
	Route 2	Route 1	16

10 Anhang III: Timetable examples for the influence analysis of HBL on operation quality

This appendix gives an example of how to statistically analyze the influence of homogeneity of blocking time on the operation quality through timetables with different value of HBL. Similar tables are also generated for the influence analysis of homogeneity of buffer time, homogeneity of running direction and overall homogeneity on operation quality.

ID	Homogeneous Interval	HBL	Delay-Coefficient	Increase of D-C	Average Increase of D-C
1		1.00	1.07	0.00	
2		0.99	1.07	0.00	
3	0.95-1.00	0.97	1.07	0.00	0.0000
4		0.96	1.07	0.00	
5		0.95	1.07	0.00	
6		0.95	1.07	0.00	
7		0.91	1.07	0.00	
8		0.90	1.14	0.07	
9		0.91	1.14	0.07	
10	0.90-0.95	0.91	1.13	0.06	0.0329
11		0.94	1.07	0.00	
12		0.90	1.09	0.02	
13		0.93	1.08	0.01	
14		0.88	1.16	0.09	
15		0.88	1.11	0.04	
16	0.85-0.90	0.85	1.15	0.08	0.0567
17		0.86	1.15	0.08	
18		0.89	1.12	0.05	
19		0.86	1.07	0.00	
20		0.82	1.07	0.00	
21		0.84	1.11	0.04	
22	0.80-0.85	0.84	1.12	0.05	0.0233
23		0.81	1.08	0.01	
24		0.83	1.10	0.03	
25		0.80	1.08	0.01	
26		0.79	1.10	0.03	
27		0.79	1.19	0.12	
28		0.79	1.07	0.00	
29		0.79	1.08	0.01	
30		0.79	1.08	0.01	
31	0.75-0.80	0.79	1.08	0.01	0.0342
32		0.78	1.20	0.13	
33		0.78	1.09	0.02	
34		0.78	1.07	0.00	
35		0.75	1.09	0.02	
36		0.75	1.11	0.04	
37		0.78	1.09	0.02	

ID	Homogeneous Interval	HBL	Delay-Coefficient	Increase of D-C	Average Increase of D-C
38		0.74	1.09	0.02	
39		0.74	1.11	0.04	
40	0.70-0.75	0.72	1.07	0.00	0.0320
41		0.71	1.10	0.03	
42		0.73	1.14	0.07	
43		0.69	1.10	0.03	
44		0.68	1.11	0.04	
45	0.65-0.70	0.68	1.09	0.02	0.0340
46		0.65	1.11	0.04	
47		0.65	1.11	0.04	
48		0.63	1.09	0.02	
49		0.63	1.12	0.05	
50		0.63	1.12	0.05	
51	0.60-0.65	0.63	1.13	0.06	0.0486
52		0.63	1.12	0.05	
53		0.63	1.12	0.05	
54		0.61	1.13	0.06	
55		0.57	1.11	0.04	
56		0.56	1.18	0.11	
57	0.55-0.60	0.58	1.22	0.15	0.0840
58		0.55	1.13	0.06	
59		0.57	1.13	0.06	
60		0.51	1.20	0.13	
61		0.53	1.13	0.06	
62	0.50-0.55	0.52	1.21	0.14	0.1200
63		0.50	1.19	0.12	
64		0.50	1.23	0.16	
65		0.50	1.18	0.11	

* D-C is the abbreviation of delay-coefficient.

11 Symbols

b_i^-	The shortest buffer time between train i and $i+1$ on a track section
dfr	The difference in free running time between train i and $i+1$ on the whole line
dsr	The difference in scheduled running time between train i and $i+1$ on the whole line
d_j	The divergence of jth parameter of homogeneity
E_{BL}	The expected value of blocking times on an occupancy element
E_{BU}	The expected value of buffer times on an occupancy element
$etBL_i$	The end time point of the blocking time of the i-th train on an occupancy element
frt_i	The free running time of ith train
HBL	The homogeneity of blocking time on an occupancy element
HBL_{IA}	The homogeneity of blocking time in an investigated area
HBU	The homogeneity of buffer time on an occupancy element
HBU_{IA}	The homogeneity of buffer time in an investigated area
h_i^-	The smallest scheduled headway between train i and $i+1$ on a track section
h_i^A	The scheduled headway at arrival point between train i and $i+1$ on a track section
h_i^D	The scheduled headway at departure point between train i and $i+1$ on a track section
h_i^{min}	The minimum headway between train i and $i+1$ on a track section
h_N	The number of headways

Hom	The overall homogeneity on an occupancy element or in an investigated area combining homogeneity of blocking time, buffer time and running time
Hom_A	The homogeneity of arrivals at a station
Hom_D	The homogeneity of departures from a station
HRD	The homogeneity of running direction on an occupancy element
HRD_{IA}	The homogeneity of running direction in an investigated area
m	the total number of occupancy elements in an investigated area
N	The number of changes in running direction
p_k	The buffer rate on occupancy element k
S_j	The entropy of jth parameter of homogeneity
srt_i	The scheduled running time of ith train
stBL_i	The start time point of the blocking time of the i-th train on an occupancy element
tBL_i	The blocking time of the i-th train on an occupancy element
tBU_j	The blocking time of the j-th buffer time on an occupancy element
V_{BL}	The variance of blocking times on an occupancy element
V_{BU}	The variance of buffer times on an occupancy element
w_k	The weight of certain occupancy element in calculating homogeneity in an investigated area
w_j	The weight for parameter of homogeneity j in calculating overall homogeneity
$y_i^{(j)}$	The value of parameter of homogeneity j for the timetable i

12   Glossary

Basic structure

According to (Martin, Li 2015), basic structure is a non-directional interrelated part of an infrastructure network. It is bounded by main signal, signal, route releasing points or the boundary of the investigated area.

Blocking time model

Blocking time model models the occupancy of the infrastructure by a train movement which is first developed by (Happel 1959).

Blocking time

The time interval in which an occupancy element is allocated to the exclusive use of one train and therefore blocked to all other trains.

Buffer time

Buffer time is the blank time interval between blocking times of two consecutive trains. For track section, the buffer time is the smallest slot between blocking time stairways of two trains. From each block section, buffer time (relate to train path) is the time gap (unoccupied time period) between two successive trains as well as intersecting trains. It is added to minimum headway to avoid a small delay transfer to the next train.

Buffer rate

Buffer rate is the sum of buffer time on an occupancy element in regard to the blocking time stairways divided by the total duration of the investigated period.

Delay-coefficient

Delay-coefficient is calculated as the ratio of out-coming delays and in-coming delays (initial delay and primary delay) through the whole network, which indicates whether in-coming delays raise or descend during the operation in the investigated area. When the value of delay-coefficient is less than 1, it indicates that the timetable is capable of eliminating some delays in this system, which can be served as a measure of the operation quality [Martin 2015].

Dispatching-related operating program parameters

The operating program parameters describe the priority of different trains and corresponding dispatching rules determine the operation quality.

Efficient infrastructure utilization	Efficient infrastructure utilization requires on one hand, a certain amount of infrastructure occupation; on the other hand, the operation quality should meet the requirement of passengers and operators.
Homogeneity in operation	Homogeneity in operation describes the homogeneity of an operating program in operation process considering external disturbances. In calculation process, the scheduled values are replaced by corresponding values with actual or theoretical data of disturbances.
Homogeneity in planning	Homogeneity in operation is the homogeneity of an operating program in planning process, which is evaluated using scheduled blocking times, scheduled buffer times and scheduled sequence of train movements.
Homogeneity of operating programs (Homogeneity of an operating program)	Homogeneity of operating programs is characterized by the variation in blocking time with respect to the occupancy of track sections, variation in buffer time with respect to headways and variation in running direction with respect to the directions of train runs, it is also called operational homogeneity.
Infrastructure occupation	Infrastructure occupation is the proportion of occupation time by all train paths on the infrastructure to the whole evaluation time window.
Interlinked occupation rate	The interlinked occupation rate is considered for the network element (e.g. line section) which consists of several single occupancy elements. On the network elements, the trains should operate without conflicts, that minimum line headway should be guaranteed between consecutive trains (DB Netz AG 2008).
Knock-on delays	A delay caused by other trains due to either short headway times or late transfer connections (Pachl 2015).
Line-related operating program parameters	The characteristics involve the operational use of a line by trains is defined as train-related operating program parameters.
Minimum line headway	The minimum line headway is the minimum time lag between two trains considering the blocking time stairways for the en-

	tire line between two trains. It can be gained through compressing the blocking time stairways of two successive trains until they touch each other without any tolerance (Pachl 2015).
Train-related operating program parameter	Train-related operating program parameters focus on the attributes of trains; including their configuration characteristics and dynamic behaviors.
Occupancy element	An occupancy element is a directed or non-directed section of exclusive accessible infrastructure, whose occupation time can be determined for each train, e.g. block section.
Occupation rate	The proportion of time occupied by train paths on an occupancy element or network elements to the whole evaluation time window.
Operating program	In railway systems, the operating program is a comprehensive date-related description of the performance and requirement of railway operation. According to DB NETZ guideline 405 (DB Netz AG 2008), an operating program is related not only to characteristics of rolling stocks themselves, but also to the structure of train runs, including the amount of train runs, the properties of trains, the structure and sequence, as well as temporal allocation of train runs.
Operating program parameter	The characteristic is to be considered in the operating program for the performance of railway operation in order to achieve the traffic demand. It was categorized into four classes: train-related, line-related, structure-related and dispatching-related operating program parameters.
Overall homogeneity	Overall homogeneity of operating programs (operational homogeneity) is calculated as the Euclidean distance of homogeneity of blocking time, homogeneity of buffer time and homogeneity of running direction to the absolute homogeneous state. It combines these three parameters of homogeneity to evaluate homogeneity of operating programs comprehensively.
Running direction	It is the direction of train movement through an occupancy element.

Single headway	Single headway is the minimum time interval between two successive trains considering for only one block section (Pachl 2015).
Single occupation rate	The proportion of time occupied by train paths on a single occupancy element to the whole evaluation time window.
Structure-related operating program parameters	Structure-related operating program parameters consider the train mix and the structure of these train movements during the whole investigated time period.

13   Bibliography

[Abril et al. 2008]   Abril, M.; Barber, F.; Ingolotti, L.; Salido, M. A.; Tormos, P.; Lova, A. (2008): An as-sessment of railway capacity. In Transportation Research Part E: Logistics and Transportation Review 44 (5), pp. 774–806. DOI: 10.1016/j.tre.2007.04.001.

[Aly et al. 2016]   Aly, M. H. F.y; Hemeda, H.; El-sayed, M. A. (2016): Computer applications in railway operation. In Alexandria Engineering Journal 55 (2), pp. 1573–1580. DOI: 10.1016/j.aej.2015.12.028.

[Bronzini and Clarke 1985]   Bronzini, M. S.; Clarke, D. B. (1985): Estimating rail line capacity and delay by computer simulation. In Tribune des Tranports 2 (1), pp. 5–11.

[Carey 1999]   Carey, M. (1999): Ex ante heuristic measures of schedule reliability. In Tranportation Research Part B 33, pp. 473–494.

[Chu 2014]   Chu, Z. (2014): Modellierung der Wartezeitfunktion bei Leistungsuntersuchungen im Schienenverkehr unter Berücksichtigung der transienten Phase. Dissertation. Universität Stuttgart, Stuttgart. Institut für Eisenbahn- und Verkehrswesen.

[DB Netz AG 2008]   DB Netz AG (2008): Fahrwegkapazität. DB 405.

[Dingler et al. 2014]   Dingler, M. H.; Lai, Y.-C.; Barkan, C. P. (2014): Effect of train-type heterogeneity on single-track heavy haul railway line capacity. In Proceedings of the Institution of Mechanical Engineers, Part F: Journal of Rail and Rapid Transit 228 (8), pp. 845–856. DOI: 10.1177/0954409713496762.

[Eichenberger 2007]   Eichenberger, P. (2007): Kapazitätssteigerung durch ETCS. In Signal und Draht 99 (3), pp. 6–14.

[Hantsch et al. 2016]   Hantsch, F.; Martin, U.; Cao, N. (2016): Effiziente Infrastrukturnutzung unter Berücksichtigung der Homogenität im Eisenbahnbetrieb. Verkehrswissenschaftliche Tage, Dresden, 2016.

[Happel 1959]   Happel, O. (1959): Sperrzeiten als grundlage für die Fahrplankonstruktion. In Eisenbahntechnische Rundschau (ETR) 8 (2), pp. 79–90.

[Hertel 1995]   Hertel, G. (1995): Leistungsfähigkeit und Leistungsverhalten von Eisenbahnbetriebsanlagen: Modelle und Berechungsmöglichkeiten. In Schriftenreihe des Instituts für Verkehrssystemtheorie und Bahnverkehr 1, pp. 62–104.

[Hertel 1992]   Hertel, Günter (1992): Die maximale Verkehrsleistung und die minimale Fahrplanempfindlichkeit auf Eisenbahnstrecken. In ETR-Eisenbahntechnische Rundschau (ETR) 10 (41), pp. 665–671.

[Kaminsky 2001]   Kaminsky, R. (2001): Pufferzeiten in Netzen des spurgeführten Verkehrs in Abhängigkeit von Zugfolge und Infrastruktur. Dissertation. Universität Hannover, Hannover. Instituts für Verkehrswesen, Eisenbahnbau und -betrieb.

[Kroon et al. 2007]   Kroon, L. G.; Dekker, R.; Vromans, M. J. C. M. (2007): Cyclic railway timetabling: A stochastic optimization approach. In F. Geraets (Ed.): Algorithmic methods for railway optimization. Lecture Notes in Computer Science, vol. 4359, Springer (2007), pp. 41-66.

[Kroon et al. 2008]   Kroon, L. G.; Maróti, G.; Helmrich, M. R.; Vromans, M.; Dekker, R. (2008). In Transportation Research Part B: Methodological 42 (6), pp. 553–570. DOI: 10.1016/j.trb.2007.11.002.

[Krueger 1999]   Krueger, H. (1999): Parametric modeling in rail capacity planning. In Farrington, P. A. Ed (Ed.): 1999 Winter simulation conference proceedings. 1999 Winter Conference on Simulation. Phoenix, AZ, USA, 5-8 Dec. 1999. Institute of Electrical and Electronics Engineers: IEEE, pp. 1194–1200.

[Lai and Barkan 2011]   Lai, Y-C.; Barkan, C. P. L. (2011): Comprehensive decision support framework for strategic railway capacity planning. In Journal Transport Engineering 137 (10), pp. 738–749. DOI: 10.1061/(ASCE)TE.1943-5436.0000248.

[Landex 2008]   Landex, A. (2008): Methods to estimate railway capacity and passenger delays. Dissertation. Kgs. Lyngby: DTU transport.

[Landex and Jensen 2013]	Landex, A.; Jensen, L. W. (2013): Measures for track complexity and robustness of operation at stations. In Journal of Rail Transport Planning & Management 3 (1-2), pp. 22–35. DOI: 10.1016/j.jrtpm.2013.10.003.
[Li 2015]	Li, X. (2015): Mikroskopische Engpassanalyse bei eisenbahnbetriebswissenschaftlichen Leistungsuntersuchungen. Dissertation. Universität Stuttgart, Stuttgart. Institut für Eisenbahn- und Verkehrswesen.
[Liang 2017]	Liang, J. (2017): Metaheuristic-based dispatching optimization integrated in multi-scale simulation model of railway operation. Dissertation. Universität Stuttgart, Stuttgart. Institut für Eisenbahn- und Verkehrswesen.
[Liang et al. 2015]	Liang, J.; Martin, U.; Cui, Y. (2015): Einfluss von ausgewählten Dispositionsparametern auf das Ergebnis von Leistungsuntersuchungen. DFG Forschungsprojekt EDELS. In Eisenbahntechnische Rundschau (ETR) 64 (07+08), pp. 20–23.
[Lindfeldt 2013]	Lindfeldt, A. (Ed.) (2013): Heterogeneity measures and secondary delays on a simulated double-track. In: Proceedings of the 5th International Seminar on Railway Operations Modelling and Analysis. Copenhagen, May 13 -15.
[Lindfeldt 2015]	Lindfeldt, A. (2015): Railway capacity analysis methods for simulation and eveluation of timetables, delays and infrastructure. Dissertation. Royal Institute of Technology (RTH), Stockholm. Department of Transport Science.
[Lindfeldt 2010]	Lindfeldt, O. (2010): Railway operation analysis. Evaluation of quality, infrastructure and timetable on single and double-track lines with analytical models and simulation. Dissertation. KTH Royal Institute of Technology, Stockholm. Department of Transport and Economics.
[Lüthi 2009]	Lüthi, M. (2009): Improving the efficiency of heavily used railway networks through integrated real-time rescheduling. Dissertation. Swiss Federal Institute of Technology, Zurich. Institute for Transport Planning and Systems.

[Martin 2015]   Martin, U. (2015): Knotenkapazität-Bewertungsverfahren für das mikroskopische Leistungsverhalten und die Engpasserkennung im spurgeführten Verkehr (RePlan). Norderstedt: BoD (Neues verkehrswissenschaftliches Journal, 8).

[Martin 2017]   Martin, U. (2017): Anforderungsgerechte Systematik zur Gestaltung von Zeitzuschlägen im Bahnbetrieb. ATRANS 1. DFG Project "MA 2326/14-1". Institut für Eisenbahn- und Verkehrswesen Stuttgart. Stuttgart.

[Martin et al. 2008]   Martin, U.; Breuer, P.; Hantschel, R. (2008): Leistungsuntersuchung Station Terminal in Stuttgart 21. unpublished. Institut für Eisenbahn- und Verkehrswesen Stuttgart. Stuttgart.

[Martin et al. 2013]   Martin, U.; Chu, Z.; Cui, Y.; Hantsch, F. (2013): Leistungsuntersuchung für große Eisenbahnknoten. Hannover, Hamburg, Mannheim, Köln, und Frankfurt am Main. unpublished. Institut für Eisenbahn- und Verkehrswesen. Stuttgart.

[Martin et al. 2011a] Martin, U.; Chu, Z.; Schmidt, C. (2011a): Leistungsuntersuchungen Weisteige Abschlussbericht. unpublished. Institut für Eisenbahn- und Verkehrswesen. Stuttgart.

[Martin and Li 2014]   Martin, U.; Li, X. (2014): Einfluss des Betriebsprogramms und der Infrastruktur auf die Entstehung von Engpässen. In Eisenbahntechnische Rundschau (ETR) 63 (3), pp. 16–21.

[Martin and Li 2015]   Martin, U.; Li, X. (2015): Entwicklung einer simulationsbasierten Methodik zur ursachenbezogenen Engpassbewertung komplexer Gleisstrukturen in spurgeführten Verkehrssystemen unter Berücksichtigung stochastischer Bedingungen. Final report of DFG Project (MA 2326/10-1). Norderstedt: BoD (Neues verkehrswissenschaftliches Journal, Ausgabe 11).

[Martin et al. 2011b] Martin, U.; Li, X.; Chu, Z. (2011b): Untersuchungsbericht zu den Anwendungsbeispielen mit PULEIV 2. im Auftrag der DB Netz AG. unpublished. Institut für Eisenbahn- und Verkehrswesen Stuttgart. Stuttgart.

[Martin et al. 2012]   Martin, U.; Li, X.; Warninghoff, C.-R. (2012): Bewertungsverfahren für Knotenelemente bei der Infrastrukturbemessung - Replan. In Eisenbahntechnische Rundschau (ETR) 11 (61), pp. 38–43.

[Martin and Liang 2017]   Martin, U.; Liang, J. (2017): The Influence of dispatching on the relationship between capacity and operation quality of railway systems. Final Report of DFG Project (MA 2326/15-1). Norderstedt: BoD (Neues verkehrswissenschaftliches Journal, Ausgabe 20).

[Pachl 2002]   Pachl, J. (2002): Railway operation and control. Mountlake Terrace WA: VTD Rail Pub.

[Pachl 2013].   Pachl, J. (2013): Systemtechnik des Schienenverkehrs. Bahnbetrieb planen, steuern und sichern. 7., überarb. u. erw. Aufl. 2014. Wiesbaden: Imprint: Springer Vieweg.

[Pachl 2015]   Pachl, J. (2015): Railway operation and control. 3rd edition. Mountlake Terrace WA: VTD Rail Pub.

[Schittenhelm 2014]   Schittenhelm, B. H. (2014): Quantitative methods for assessment of railway timetables. Dissertation. Technical University of Denmark, Denmark. Department of Transport.

[Transportation Research Board 2013]   Transportation Research Board (2013): Transit capacity and quality of service manual. Chapter 8: Rail Transit Capacity. 3rd. Edited by Transportation Research Board. Washington, DC.

[UIC 2004]   UIC (2004): Capacity (1st edition). UIC Code 406.

[UIC 2013]   UIC (2013): Capacity (2nd edition). UIC Code 406.

[Vromans 2005]   Vromans, M. J.C.M. (2005): Reliability of railway systems. Dissertation. Erasmus University Rotterdam, Rotterdam. Institute of Management (ERIM).

[Vromans et al. 2006]   Vromans, M. J.C.M.; Dekker, R.; Kroon, L. G. (2006): Reliability and heterogeneity of railway services. In European Journal of Operational Research 172 (2), pp. 647–665. DOI: 10.1016/j.ejor.2004.10.010.

[Wendler 2007]   Wendler, E. (2007): The scheduled waiting time on railway lines. In Transportation Research Part B: Methodological 41 (2), pp. 148–158. DOI: 10.1016/j.trb.2006.02.009.

[Yuan and Hansen 2007] Yuan, J.; Hansen, I. A. (2007): Optimizing capacity utilization of stations by estimating knock-on train delays. In Transportation Research Part B: Methodological 41 (2), pp. 202–217. DOI: 10.1016/j.trb.2006.02.004.